T0318723

LEARNING-BASED ADAPTIVE CONTROL

LEARNING-BASED ADAPTIVE CONTROL

An Extremum Seeking Approach - Theory and Applications

MOUHACINE BENOSMAN

AMSTERDAM • BOSTON • HEIDELBERG • LONDON
NEW YORK • OXFORD • PARIS • SAN DIEGO
SAN FRANCISCO • SINGAPORE • SYDNEY • TOKYO

Butterworth-Heinemann is an imprint of Elsevier

Butterworth-Heinemann is an imprint of Elsevier
The Boulevard, Langford Lane, Kidlington, Oxford OX5 1GB, UK
50 Hampshire Street, 5th Floor, Cambridge, MA 02139, USA

Library of Congress Cataloging-in-Publication Data
A catalog record for this book is available from the Library of Congress

British Library Cataloguing-in-Publication Data
A catalogue record for this book is available from the British Library

ISBN: 978-0-12-803136-0

For information on all Butterworth Heinemann publications
visit our website at https://www.elsevier.com/

Working together
to grow libraries in
developing countries

www.elsevier.com • www.bookaid.org

Publisher: Joe Hayton
Acquisition Editor: Sonnini R. Yura
Editorial Project Manager: Ana Claudia Abad Garcia
Production Project Manager: Kiruthika Govindaraju
Cover Designer: Victoria Pearson Esser

Typeset by SPi Global, India

CONTENTS

PREFACE

Paris 18th, 1914: The day was, by all accounts, a sunny and pleasant one, the blue sky a perfect backdrop for spectacle. A large crowd had gathered along the banks of the Seine, near the Argenteuil bridge in the city's northwestern fringes, to witness the Concours de la Sécurité en Aéroplane, an aviation competition organized to show off the latest advances in flight safety. Nearly sixty planes and pilots took part, demonstrating an impressive assortment of techniques and equipment. Last on day's program, flying a Curtiss C-2 biplane, was a handsome American pilot named Lawrence Sperry. Sitting beside him in the C-2's open cockpit was his French mechanic, Emil Cachin. As Sperry flew past the ranks of spectators and approached the judges' stand, he let go of the plane's controls and raised his hands. The crowd roared. The plane was flying itself!

Carr (2014, Chapter 3)

This was, a hundred years ago, one of the first demonstrations of adaptation and control implemented in hardware at that time. Since then, adaptive control has been one of the main problems studied in control theory. The problem is well understood, yet it has a very active research frontier (cf. Chapter 2). This monograph focuses on a specific subclass of adaptive control, namely learning-based adaptive control.

As we will see in Chapter 2, adaptive control can be divided into three main subclasses: the classical model-based adaptive control, which mainly uses physics-based models of the controlled system; the model-free adaptive control, which is solely based on the interaction of the controller with the system; and learning-based adaptive control, which uses both model-based and model-free techniques to design flexible yet fast and stable (i.e., safe) adaptive controllers. The basic idea of learning-based modular adaptive control that we introduce in this monograph is depicted in Fig. A. We see that there are two main blocks: the model-based block and the model-free block. The model-based part is concerned with ensuring some type of stability during the learning process. The model-free (i.e., learning) part is concerned with improving the performance of the controllers by tuning online some parameters of the model-based controllers. Due to the modular design, the two blocks can be connected safely, that is, without jeopardizing the stability (in the sense of boundedness) of the whole system.

We argue that one of the main advantages of this type of adaptive controllers, compared to other approaches of adaptation, is the fact that they ensure stability of the system, yet they take advantage of the flexibility of model-free learning algorithms. Model-based adaptive controllers can be

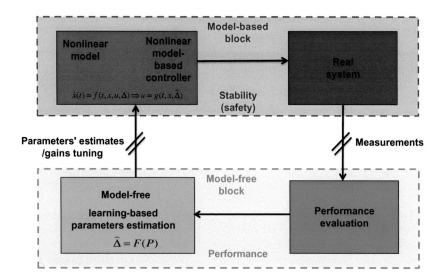

Fig. A Block diagram of the modular learning-based adaptive control.

very efficient and stable. However, they impose many constraints on the model, as well as on the uncertainties' structure (e.g., linear vs nonlinear structures, etc.). Model-free adaptive controllers, on the other hand, allow a lot of flexibility in terms of model structures because they do not rely on any model. However, they lack some of the stability guarantees which characterize model-based adaptive controllers. Furthermore, model-free adaptive algorithms have to learn the best control action (or policy) over a large domain of control options because they do not take advantage of any physical knowledge of the system; that is, they do not use any model of the system. Learning-based adaptive controllers strike a balance; they have some of the stability guarantees due to their model-based part, but also are expected to converge faster to the optimal performance, compared to their model-free counterparts. This is due to the fact that they use some initial knowledge and modeling of the system, albeit uncertain or incomplete.

To make the book easier to read, without the need to refer to other sources, we will recall in Chapter 1 the main definitions and tools used in control theory. This includes the classical definitions of vector spaces, Hilbert spaces, invariant sets, and so on. Other useful stability concepts like Lyapunov, Lagrange stability, and input-to-state stability (ISS) will also be recalled. Finally, some important notions of passivity and nonminimum phase will be presented as well.

In Chapter 2, we will present a general survey of the adaptive control field. We will classify some of the main relevant results from a few subfields of adaptive control theory. The main goal of this chapter is to situate the results of this monograph in the global picture of adaptive control, so that the reader can better understand where our results stand, and how they differ from other works.

The remaining chapters are more technical because they are about more specific results of learning-based adaptation, which we have been working on for the past 5 years.

Starting with Chapter 3, we focus on a very specific problem in learning-based adaptation, namely the problem of iterative feedback tuning (IFT). The main goal of IFT is to automate the tuning of feedback gains for linear or nonlinear feedback controllers. We will first give a brief overview of some IFT research results, and then introduce our work in this field. More specifically, we will focus on extremum seeking-based nonlinear IFT. Throughout this book, we will often focus on nonlinear models (maybe with the exception of Chapter 6) because we believe that, with some simplifications, the nonlinear results can easily be applied to linear models. We will not, however, explicitly derive such simplifications; we leave it to the interested reader to apply the techniques presented here to linear plants.

Chapter 4 presents the general formulation of our modular extremum seeking-based adaptive control for nonlinear models. We will start this chapter with the case of general nonlinear models, without imposing any structural constraints on the model's equations or on the uncertainties (besides some basic smoothness constraints). For this rather general class of models we will argue that, under the assumption of input-to-state stabilizability (by feedback), we can design modular learning-based indirect adaptive controllers, where a model-free learning algorithm is used to estimate online the model's parametric uncertainties. We then focus on a more specific class of nonlinear systems, namely nonlinear systems affine in the control vector. For this class of nonlinear systems we present a constructive control design, which ensures the ISS, and then complement it with an extremum seeking model-free learning algorithm to estimate the model's uncertainties.

In Chapter 5, we will study the problem of real-time nonlinear model identification. What we mean by real time is that we want to identify some parameters of the system online while the system is performing its nominal tasks, without the need to stop or change the system's tasks solely for the purpose of identification. Indeed, real-time identification, if achieved properly, can be of great interest in industrial applications, where interrupting a system's task can lead to big financial losses. If we can identify the system's parameters and keep updating them in real time, we can track their drift, for instance due to the aging of the system or due to a change in the nominal task (a manipulator arm moving different parts, with different masses, etc.) in real time, and then update the model accordingly.

We will study the problem of extremum seeking-based parametric model identification for both finite dimension ordinary differential equation models and infinite dimension partial differential equations (PDEs). We will also study in Chapter 5 a related problem, namely reduced order stabilization for PDEs. In this problem we will use model-free extremum seekers to auto-tune stabilizing terms, known as closure models, which are used to stabilize reduced order models obtained by projecting the PDEs onto a finite dimensional space.

Finally, as a by-product of the approach advocated in this book, we will study in Chapter 6 the specific problem of model predictive control (MPC) for linear models with parametric uncertainties. This case can be seen as a special case of the general results presented in Chapter 4, where the controllers ensuring ISS, are in the form of a model predictive controller. We will use recent results in the field to design an MPC with ISS properties (which is a rather standard result, due to the numerous recent papers on the ISS-MPC topic), and then *carefully and properly* complement it with a model-free extremum seeker to iteratively learn the model's uncertainties and improve the overall performance of the MPC.

"Conclusions and Further Notes" chapter summarizes the presented results. We close with a few thoughts about possible extensions of the results of this book, and mention some open problems which we think are important to investigate in future adaptive control research.

M. Benosman
Cambridge, MA, United States
March 2016

REFERENCES

N. Carr, The Glass Cage: How Our Computers Are Changing Us, W.W. Norton & Company Ltd., New York, 2014.

ACKNOWLEDGMENTS

I would like to thank everyone who has supported me along the way, including my family, collaborators, and coworkers—in particular, Anthony Vetro, Petros Boufonous, Arvind Raghunathan, Daniel Nikovski, Abraham Goldsmith, and Piyush Grover. I would also like to thank Matthew Brand and Alan Sullivan for proofreading the opening and closing chapters of this book, and for their friendship and support. My gratitude also goes to all the students with whom I have worked over the past 5 years; some parts of the research presented in this book are based on collaborations with Gokhan Atinc, Anantharaman Subbaraman, Meng Xia, and Boris Kramer. A special thanks to my friends, Jennifer Milner and Marjorie Berkowitz, for their encouragement. Finally, I want to thank my editors for their patience throughout the course of this project.

As a special note, I would also like to dedicate this modest work to my late colleague, Mr John C. Barnwell III, a genius hardware expert, and most important, a genuine person, who left us prematurely, while I was finishing this book.

Dove si urla, non c'è vera conoscenza. (Where there is shouting, there is no true knowledge.)

Leonardo da Vinci

Mon verre n'est pas grand mais je bois dans mon verre. (My glass is not large, but I drink from my glass.)

Alfred Louis Charles de Musset

CHAPTER 1

Some Mathematical Tools

We will report here most of the mathematical tools which will be used throughout this book. The goal is to give the reader the main tools to be able to understand the remaining chapters of this monograph. The concepts presented here might seem too general because the chapter includes all the mathematical definitions which will be used at some point in the book. However, later on, to make it more specific, in each of the chapters we will start with a brief recall of the main mathematical tools which will be needed in each specific chapter.

We start with some useful definitions and properties related to vectors, matrices, and functions, see Golub and Van Loan (1996).

1.1 NORMS DEFINITIONS AND PROPERTIES

Let us start by recalling the important definition of vector spaces, which are often used in control theory.

Definition 1.1. *A vector space is a set V on which two operations $+$ and $\cdot$ are defined, called vector addition and scalar multiplication.*

The operation $+$ (vector addition) must satisfy the following conditions:
- *Closure: if u and v are any vectors in V, then the sum $u + v$ belongs to V.*
- *Commutative law: for all vectors u and v in V, $u + v = v + u$.*
- *Associative law: for all vectors u, v, and w in V, $u + (v + w) = (u + v) + w$.*
- *Additive identity: the set V contains an additive identity element, denoted by 0, such that for any vector v in V, $0 + v = v$ and $v + 0 = v$.*
- *Additive inverses: for each vector v in V, the equations $v + x = 0$ and $x + v = 0$ have a solution x in V, called an additive inverse of v, and denoted by $-v$.*

The operation $\cdot$ (scalar multiplication) is defined between real numbers (or scalars) and vectors, and must satisfy the following conditions:
- *Closure: if v is any vector in V, and c is any real number, then the product $c \cdot v$ belongs to V.*
- *Distributive law: for all real numbers c and all vectors u, v in V, $c \cdot (u + v) = c \cdot u + c \cdot v$.*

Learning-Based Adaptive Control
http://dx.doi.org/10.1016/B978-0-12-803136-0.00001-4

- *Distributive law: for all real numbers c, d and all vectors v in V, $(c + d) \cdot v = c \cdot v + d \cdot v$.*
- *Associative law: for all real numbers c, d and all vectors v in V, $c \cdot (d \cdot v) = (cd) \cdot v$.*
- *Unitary law: for all vectors v in V, $1 \cdot v = v$.*

A very well-known example of vector spaces is the space $\mathbb{R}^n$, $n \in \mathbb{Z}^+$. It will often be used in this book to define our mathematical models.

Definition 1.2. *A Hilbert space is a vector space $\mathcal{H}$ associated with an inner product $\langle ., . \rangle$ such that the norm defined by $|f|^2 = \langle f, f \rangle$, $\forall f \in \mathcal{H}$, makes $\mathcal{H}$ a complete metric space.*

A well-known example of finite-dimension Hilbert spaces is the $\mathbb{R}^n$ associated with the inner product $\langle u, v \rangle = u^T v$, that is, the vector dot product.

Next, we recall the definition of norms of a vector x in a vector space $\mathbb{V}$. The norm of x can be seen as the extension to vector objects of the classical absolute value for a scalar element $|a|$, $a \in \mathbb{R}$. A more rigorous definition is given next.

Definition 1.3. *The norm of a vector $x \in \mathbb{V}$ is a real valued function $\|.\|$: $\mathbb{V} \to \mathbb{R}$, s.t.,*

1. $\|x\| \geq 0$, *with* $\|x\| = 0$ *iff* $x = 0$
2. $\|ax\| = |a| \|x\|$, $\forall a \in \mathbb{R}$
3. $\|x + y\| \leq \|x\| + \|y\|$, $\forall y \in \mathbb{V}$

Some examples of vector norms are as follows:

- *The infinity norm:* $\|x\|_\infty = \max_i |x_i|$
- *The one norm:* $\|x\|_1 = \sum_i |x_i|$
- *The Euclidean (or two) norm:* $\|x\|_2 = \sqrt{\sum_i |x_i|^2}$
- *The p-norm:* $\|x\|_p = \left(\sum_i |x_i|^p \right)^{1/p}$, $\forall p \in \mathbb{Z}^+$

We recall that these norms are equivalent in $\mathbb{R}^n$, that is, $\forall x \in \mathbb{R}^n$; these norms satisfy the inequalities

$$\begin{aligned} \|x\|_\infty &\leq \|x\|_2 \leq \sqrt{n} \|x\|_\infty, \\ \|x\|_\infty &\leq \|x\|_1 \leq n \|x\|_\infty. \end{aligned} \tag{1.1}$$

The Euclidean norm satisfies the following (Cauchy-Schwarz) inequality

$$|x^T y| \leq \|x\|_2 \|y\|_2, \quad \forall x, y \in \mathbb{R}^n. \tag{1.2}$$

We define now the induced matrix norms associated with vector norms. We recall that a matrix $A \in \mathbb{R}^{m \times n}$ consists of m rows, and n columns of real elements. It can be defined as the linear operator $A(.) : \mathbb{R}^n \to \mathbb{R}^m$, s.t.,

$$\tilde{x} = Ax, \ \tilde{x} \in \mathbb{R}^m, \ x \in \mathbb{R}^n. \tag{1.3}$$

Definition 1.4. *For a given vector norm* $\|x\|$, $\forall x \in \mathbb{R}^n$, *we define the induced matrix norm* $\|A\|$, $\forall A \in \mathbb{R}^{m \times n}$, *as*

$$\|A\| = \sup_{x \neq 0} \frac{\|Ax\|}{\|x\|} = \sup_{\|x\|=1} \|Ax\|, \quad \forall x \in \mathbb{R}^n. \tag{1.4}$$

Some properties of induced matrix norms are
1. $\|Ax\| \leq \|A\|\|x\|, \quad \forall x \in \mathbb{R}^n$
2. $\|A + B\| \leq \|A\| + \|B\|$
3. $\|AB\| \leq \|A\| + \|B\|$

Examples of induced norms for $A \in \mathbb{R}^{m \times n}$ are
- *The infinity induced matrix norm:* $\|A\|_\infty = \max_i \sum_j |a_{ij}|$
- *The one induced matrix norm:* $\|A\|_1 = \max_j \sum_i |a_{ij}|$
- *The two induced matrix norm:* $\|A\|_2 = \sqrt{\lambda_{\max}(A^T A)}$, where $\lambda_{\max}(A)$ is the maximum eigenvalue of A

Another frequently used matrix norm is the Frobenius norm, defined as $\|A\|_F = \sqrt{\sum_i \sum_j |a_{ij}|^2}$. It is worth noting that the Frobenius norm is not induced by any vector p-norm.

Some other useful definitions and properties related to matrices are recalled below.

Definition 1.5. *A real symmetric matrix* $A \in \mathbb{R}^{n \times n}$ *is positive (negative) semidefinite if* $x^T Ax \geq 0$ $(x^T Ax \leq 0)$, *for all* $x \neq 0$.

Definition 1.6. *A real symmetric matrix* $A \in \mathbb{R}^{n \times n}$ *is positive (negative) definite if* $x^T Ax > 0$ $(x^T Ax < 0)$, *for all* $x \neq 0$.

Definition 1.7. *The leading principles of a real symmetric matrix* $A \in \mathbb{R}^{n \times n}$ *are defined as the submatrices* $A_i = \begin{bmatrix} a_{11} & \cdots & a_{1i} \\ \vdots & & \vdots \\ a_{i1} & \cdots & a_{ii} \end{bmatrix}$, $i = 2, \ldots, n$.

The positive (negative) definiteness of a matrix can be tested by using the following properties:
1. A real symmetric matrix $A \in \mathbb{R}^{n \times n}$ is positive (negative) definite iff $\lambda(A)_i > 0$ $(\lambda(A)_i < 0)$, $\forall i = 1, \ldots, n$.
2. A real symmetric matrix $A \in \mathbb{R}^{n \times n}$ is positive (negative) definite iff all the determinants of its leading principles $\det(A_i)$ are positive (negative).

Other useful properties are as follows:
1. If $A > 0$, then its inverse A^{-1} exists and is positive definite.
2. If $A_1 > 0$, and $A_2 > 0$, then $\alpha A_1 + \beta A_2 > 0$ for all $\alpha > 0$, $\beta > 0$.

3. If $A \in \mathbb{R}^{n \times n}$ is positive definite, and $C \in \mathbb{R}^{m \times n}$ is of rank m, then $B = CAC^T \in \mathbb{R}^{m \times m}$ is positive definite.
4. The Rayleigh-Ritz inequality: for any real symmetric matrix $A \in \mathbb{R}^{n \times n}$, $\lambda(A)_{min} x^T x \le x^T A x \le \lambda(A)_{max} x^T x$, $\forall x \in \mathbb{R}^n$.

The norms and properties defined so far are for constant objects, that is, constant vectors and matrices. It is possible to extend some of these definitions to the case of time-varying objects, that is, time-varying vector functions, and matrices. We recall some of these definitions next.

Consider a vector function $x(t) : \mathbb{R}^+ \to \mathbb{R}^n$, with $\mathbb{R}^+ = [0, \infty[$, its p-norm is defined as

$$\|x(t)\|_p = \left(\int_0^\infty \|x(\tau)\|^p d\tau \right)^{1/p}, \tag{1.5}$$

where $p \in [1, \infty[$, and the norm inside the integral can be any vector norm in $\mathbb{R}^n$. The function is said to be in $\mathcal{L}_p$ if its p-norm is finite. Furthermore, the infinity-norm in this case is defined as

$$\|x(t)\|_\infty = \sup_{t \in \mathbb{R}^+} \|x(t)\|. \tag{1.6}$$

The function is said to be in $\mathcal{L}_\infty$ if its infinity-norm is finite.

Some properties of these function norms are given below:
1. Hölder's inequality: for p, $q \in [1, \infty[$, such that $\frac{1}{p} + \frac{1}{q} = 1$, if $f \in \mathcal{L}_p$, $g \in \mathcal{L}_q$, then $\|fg\|_1 \le \|f\|_p \|g\|_q$.
2. Schwartz inequality: for $f, g \in \mathcal{L}_2$, $\|fg\|_1 \le \|f\|_2 \|g\|_2$.
3. Minkowski inequality: for $f, g \in \mathcal{L}_p$, $p \in [1, \infty[$, $\|f + g\|_p \le \|f\|_p + \|g\|_p$.

Next, to slowly move toward the definitions related to dynamical systems and their stability, we introduce some basic functions' definitions and properties, see Perko (1996) and Khalil (1996).

1.2 VECTOR FUNCTIONS AND THEIR PROPERTIES

We will recall here some basic definitions of functions' properties which are often used in dynamical systems theory.

Definition 1.8. *A function $f : [0, \infty[\to \mathbb{R}$ is continuous on $[0, \infty[$, if for any $\epsilon > 0$ there exists a $\delta(t, \epsilon)$, such that for any t, $\tilde{t} \in [0, \infty[$, with $|t - \tilde{t}| < \delta(t, \epsilon)$ we have $|f(t) - f(\tilde{t})| < \epsilon$.*

Definition 1.9. *A function $f : [0, \infty[\to \mathbb{R}$ is uniformly continuous on $[0, \infty[$, if for any $\epsilon > 0$ there exists a $\delta(\epsilon)$, such that for any t, $\tilde{t} \in [0, \infty[$, with $|t - \tilde{t}| < \delta(\epsilon)$ we have $|f(t) - f(\tilde{t})| < \epsilon$.*

Definition 1.10. *A function $f : [0, \infty[\to \mathbb{R}$ is piecewise continuous on $[0, \infty[$, if f is continuous on a finite number of intervals $]t_i, t_{i+1}[\subset [0, \infty[$, $i = 1, 2, \ldots$, and discontinuous at a finite number of points between these intervals.*

Definition 1.11. *A vector function $f : \mathbb{R}^n \to \mathbb{R}^m$ $(n, m = 1, 2, \ldots)$ is uniformly k_l-Lipschitz continuous (or k_l-Lipschitz, for short) if there exists a positive (Lipschitz) constant k_l, s.t., $\|f(x) - f(y)\| \leq k_l \|x - y\|$, $\forall x, y \in \mathbb{R}^n$.*

Of course, Definition 1.11 readily extends to the case where f is defined on a subset $\mathcal{D} \subset \mathbb{R}^n$, in which case the function is said to be Lipschitz on D.

Definition 1.12. *A vector function $f : \mathbb{R}^n \to \mathbb{R}^m$ $(n, m = 1, 2, \ldots)$ is said to be radially unbounded (also referred to as proper in some literature) if $\|f(x)\| \to \infty$ for $\|x\| \to \infty$.*

Definition 1.13. *A scalar function $f : \mathbb{R}^n \to \mathbb{R}$ $(n = 1, 2, \ldots)$ is said to be positive (negative) definite if $f(x) > 0$ $(f(x) < 0)$, $\forall x \neq 0$.*

Definition 1.14. *A scalar function $f : \mathbb{R}^n \to \mathbb{R}$ $(n = 1, 2, \ldots)$ is said to be positive (negative) semidefinite if $f(x) \geq 0$ $(f(x) \leq 0)$, $\forall x \neq 0$.*

Definition 1.15. *A scalar function $f : \mathbb{R}^n \to \mathbb{R}$ $(n = 1, 2, \ldots)$ is said to be C^k smooth on $\mathbb{R}^n$, if all the functions $f^{(k)}$, $k = 0, 1, 2, \ldots, k$, are continuous on $\mathbb{R}^n$. If $k \to \infty$, we simply say that f is smooth.*

One important result related to continuous functions is the well-known Barbalat's lemma, see Khalil (1996), which is very often used in control theory when dealing with nonautonomous systems (i.e., models with explicit time variable).

Lemma 1.1 (Barbalat's lemma). *Consider a function $f : \mathbb{R}^+ \to \mathbb{R}$ of the variable $t \in \mathbb{R}^+$, such that f is uniformly continuous on $\mathbb{R}^+$, if $\lim_{t \to \infty} \int_0^T f(\tau) d\tau$ exists and is finite, then $\lim_{t \to \infty} f(t) = 0$.*

A direct consequence of Barbalat's lemma is stated next.

Lemma 1.2. *Consider a function $f : \mathbb{R}^+ \to \mathbb{R}$ of the variable $t \in \mathbb{R}^+$, if $f \in \mathcal{L}_\infty, \dot{f} \in \mathcal{L}_\infty$, and $f \in \mathcal{L}_p$ for a given $p \in [1, \infty[$, then $f(t) \to 0$.*

Another important definition, especially in the context of adaptive control, is the well-known persistent excitation (PE) condition for a function.

Definition 1.16. *The vector function $f : [0, \infty) \to \mathbb{R}^m$ is said to satisfy the PE condition on $[0, \infty)$ if it is measurable and bounded on $[0, \infty)$, and there exist $\alpha > 0$, $T > 0$, such that*

$$\int_t^{t+T} f(\tau) f(\tau)^T d\tau \geq \alpha I_m, \quad \forall t \geq 0, \tag{1.7}$$

where I_m denotes the identity matrix of dimension $m \times m$.

We finalize this section with a few basic definitions which are very useful in control theory.

Definition 1.17. *The Jacobian matrix of a vector function* $f : \begin{smallmatrix} \mathbb{R}^n \to \mathbb{R}^m \\ x \to f(x) \end{smallmatrix}$ *is defined as*

$$
\frac{\partial f}{\partial x} = \begin{bmatrix} \frac{\partial f_1}{\partial x_1} & \cdots & \frac{\partial f_1}{\partial x_n} \\ \vdots & & \vdots \\ \frac{\partial f_m}{\partial x_1} & \cdots & \frac{\partial f_m}{\partial x_n} \end{bmatrix} \tag{1.8}
$$

Definition 1.18. *The gradient vector of a scalar function* $f : \begin{smallmatrix} \mathbb{R}^n \to \mathbb{R} \\ x \to f(x) \end{smallmatrix}$ *is defined as*

$$
\nabla f_x = \left[\frac{\partial f}{\partial x_1}, \ldots, \frac{\partial f}{\partial x_n} \right]^T \in \mathbb{R}^n \tag{1.9}
$$

Definition 1.19. *The Hessian matrix of a scalar function is defined as the Jacobian of its gradient vector function.*

Definition 1.20. *The Lie derivative of a scalar function* $f : \begin{smallmatrix} \mathbb{R}^n \to \mathbb{R} \\ x \to f(x) \end{smallmatrix}$ *along a vector function* $g : \begin{smallmatrix} \mathbb{R}^n \to \mathbb{R}^n \\ x \to g(x) \end{smallmatrix}$ *is defined as* $L_g f = \nabla f_x^T g(x)$. *A high order Lie derivative is defined as* $L_g^k f = L_g L_g^{k-1} f$, $k \geq 1$, *with* $L_g^0 f = f$.

Definition 1.21. *A function is said to be analytic in a given set if it admits a convergent Taylor series approximation in some neighborhood of every point of the set.*

We take this opportunity to amend a common confusion which we encountered over and over again in the control literature, where some authors refer to analytic function when they want to emphasize that the function (usually a controller) is derived as a symbolic expression (in other words, not in a numerical form). The right way to characterize such functions is to say that the function (or control) is given in closed form (whereas analytic should be used as defined previously). At least in this monograph, we shall distinguish between these two terms.

Definition 1.22. *A continuous function* $\alpha : [0, a) \to [0, \infty)$ *is said to belong to class* $\mathcal{K}$ *if it is strictly increasing and* $\alpha(0) = 0$.

Definition 1.23. *A continuous function* $\beta : [0, a) \times [0, \infty) \to [0, \infty)$ *is said to belong to class* $\mathcal{KL}$ *if, for each fixed* s, *the mapping* $\beta(r, s)$ *belongs to class* $\mathcal{K}$ *with respect to* r *and, for each fixed* r, *the mapping* $\beta(r, s)$ *is decreasing with respect to* s *and* $\beta(r, s) \to 0$ *as* $s \to \infty$.

Let us now introduce some stability definitions for dynamical systems, see Liapounoff (1949), Rouche et al. (1977), Lyapunov (1992), and Zubov (1964).

1.3 STABILITY OF DYNAMICAL SYSTEMS

To introduce some stability concepts, we consider a general dynamical time-varying system of the form

$$\dot{x}(t) = f(t, x(t)), \quad x(t_0) = x_0, \quad t \geq t_0 \geq 0 \tag{1.10}$$

where $x(t) \in \mathcal{D} \subseteq \mathbb{R}^n$ such that $0 \in \mathcal{D}$, $f : [t_0, t_1) \times \mathcal{D} \to \mathbb{R}^n$ is such that $f(\cdot, \cdot)$ is jointly continuous in t and x, and for every $t \in [t_0, t_1)$, $f(t, 0) = 0$, and $f(t, \cdot)$ is locally Lipschitz in x uniformly in t for all t in compact subsets of $[0, \infty)$. These assumptions guarantee the existence and uniqueness of the solution $x(t, t_0, x_0)$ over the interval $[t_0, t_1)$.

Definition 1.24. *A point $x_e \in \mathbb{R}^n$ is an equilibrium point for the system* (1.10) *if $f(t, x_e) = 0$, $\forall t \geq t_0$.*

Definition 1.25 (Stability in the sense of Lyapunov). *An equilibrium point x_e of the system* (1.10) *is said to be stable if for any $t_0 \geq 0$, and $\epsilon > 0$, there exists a $\delta(t_0, \epsilon)$ such that*

$$\|x_0 - x_e\| < \delta(t_0, \epsilon) \Rightarrow \|x(t, t_0, x_0) - x_e\| < \epsilon, \quad \forall t \geq t_0.$$

Definition 1.26 (Uniform stability in the sense of Lyapunov). *If the parameter δ in the stability Definition 1.25 is independent of time, that is, independent of t_0, then the equilibrium point x_e is said to be uniformly stable.*

Definition 1.27 (Asymptotic stability in the sense of Lyapunov). *An equilibrium point x_e of the system* (1.10) *is said to be asymptotically stable if it is stable and there exists a positive constant $c(t_0)$, such that $x(t) \to x_e$, for all $\|x(t_0) - x_e\| < c$.* (as $t \to \infty$)

Definition 1.28 (Uniform asymptotic stability in the sense of Lyapunov). *An equilibrium point x_e of the system* (1.10) *is said to be uniformly asymptotically stable if it is uniformly stable and there exists a positive constant c, such that $x(t) \to x_e$, for all $\|x(t_0) - x_e\| < c$ uniformly in t_0, that is, for each $\alpha > 0$, $\exists T(\alpha) > 0$, s.t. $\|x(t) - x_e\| < \alpha$, $\forall t \geq t_0 + T(\alpha)$, $\forall \|x(t_0) - x_e\| < c$.* (as $t \to \infty$)

Definition 1.29 (Globally uniform asymptotic stability in the sense of Lyapunov). *An equilibrium point x_e of the system* (1.10) *is said to be globally uniformly asymptotically stable if it is uniformly stable with $\lim_{\epsilon \to \infty} \delta(\epsilon) = \infty$, and* *there exists a positive constant $T(\alpha, \beta) > 0$ for each pair $(c_1, c_2) \in \mathbb{R}^+ \times \mathbb{R}^+$, such that $\|x(t) - x_e\| < \alpha$, $\forall t \geq t_0 + T(c_1, c_2)$, $\forall \|x(t_0) - x_e\| < c$.*

Definition 1.30 (Exponential stability). *An equilibrium point x_e of the system (1.10) is said to be exponentially stable if there exist positive constants c_1, c_2, and c_3 such that*

$$\|x(t) - x_e\| \leq c_1 \|x(t_0) - x_e\| e^{-c_2(t-t_0)}, \quad \forall t \geq t_0 \geq 0, \quad \forall \|x(t_0) - x_e\| < c_3. \tag{1.11}$$

Moreover, the stability is global if the first inequality in Eq. (1.11) holds for any $x(t_0)$.

We can also recall here an important lemma which characterizes uniform asymptotic stability in terms of class $\mathcal{K}$ and class $\mathcal{KL}$ functions, see for example, Khalil (1996).

Lemma 1.3. *An equilibrium point x_e of the system (1.10) is*

1. *Uniformly stable if and only if there exist a class $\mathcal{K}$ function α, and a positive constant c independent of t_0, such that*

$$\|x(t) - x_e\| \leq \alpha(\|x(t_0) - x_e\|), \quad \forall t \geq t_0 \geq 0, \quad \forall \|x(t_0) - x_e\| < c \tag{1.12}$$

2. *Uniformly asymptotically stable if and only if there exist a class $\mathcal{KL}$ function β and a positive constant c, independent of t_0, such that*

$$\|x(t) - x_e\| \leq \beta(\|x(t_0) - x_e\|, t - t_0), \quad \forall t \geq t_0 \geq 0, \quad \forall \|x(t_0) - x_e\| < c \tag{1.13}$$

3. *Globally uniformly asymptotically stable, if and only if condition (2) of this lemma is satisfied for any initial state $x(t_0)$.*

Definition 1.31 ((ϵ, δ)-Semiglobal practical uniform ultimate boundedness—SPUUB). *Consider the system*

$$\dot{x} = f^\epsilon(t, x) \tag{1.14}$$

with $\phi^\epsilon(t, t_0, x_0)$ being the solution of Eq. (1.14) starting from the initial condition $x(t_0) = x_0$. Then, the origin of Eq. (1.14) is said to be (ϵ, δ)-SPUUB if it satisfies the following three conditions:

1. *(ϵ, δ)-Uniform stability: For every $c_2 \in]\delta, \infty[$ there exists $c_1 \in]0, \infty[$ and $\hat{\epsilon} \in]0, \infty[$ such that for all $t_0 \in \mathbb{R}$ and for all $x_0 \in \mathbb{R}^n$ with $\|x_0\| < c_1$ and for all $\epsilon \in]0, \hat{\epsilon}[$,*

$$\|\phi^\epsilon(t, t_0, x_0)\| < c_2, \quad \forall t \in [t_0, \infty[$$

2. *(ϵ, δ)-Uniform ultimate boundedness: For every $c_1 \in]0, \infty[$ there exists $c_2 \in]\delta, \infty[$ and $\hat{\epsilon} \in]0, \infty[$ such that for all $t_0 \in \mathbb{R}$ and for all $x_0 \in \mathbb{R}^n$ with*

$\|x_0\| < c_1$ and for all $\epsilon \in]0, \hat{\epsilon}[$,

$$\|\phi^\epsilon(t, t_0, x_0)\| < c_2, \quad \forall t \in [t_0, \infty[$$

3. (ϵ, δ)-*Global uniform attractivity: For all c_1, $c_2 \in (\delta, \infty)$ there exists $T \in$ $]0, \infty[$ and $\hat{\epsilon} \in]0, \infty[$ such that for all $t_0 \in \mathbb{R}$ and for all $x_0 \in \mathbb{R}^n$ with $\|x_0\| < c_1$ and for all $\epsilon \in]0, \hat{\epsilon}[$,*

$$\|\phi^\epsilon(t, t_0, x_0)\| < c_2, \quad \forall t \in [t_0 + T, \infty[$$

Based on the previous definitions we will present now some stability theorems, where we assume without loss of generality that the equilibrium point has been shifted to the origin by a simple change of coordinates.

Theorem 1.1 (Lyapunov's direct method- Stability theorem). *Let us assume that $x_e = 0$ is an equilibrium point for the system dynamics (1.10), and $\mathcal{D} \subset \mathbb{R}^n$ is a domain containing $x = 0$. Let $V : [0, \infty[\times \mathcal{D} \to \mathbb{R}$ be a continuously differentiable positive definite function such that*

$$W_1(x) \le V(t, x) \le W_2(x), \tag{1.15}$$

$$\frac{\partial V}{\partial t} + \frac{\partial V}{\partial x} f(t, x) \le 0, \quad \forall t \ge 0, \quad \forall x \in \mathbb{D}, \tag{1.16}$$

where $W_1(x)$ and $W_2(x)$ are continuous positive definite functions on $\mathcal{D}$. Then, the origin is uniformly stable.

Theorem 1.2 (Lyapunov's direct method—Asymptotic stability theorem). *Consider the same statement of Theorem 1.1, where the condition (1.16) is replaced with*

$$\frac{\partial V}{\partial t} + \frac{\partial V}{\partial x} f(t, x) \le -W_3(x), \quad \forall t \ge 0, \quad \forall x \in \mathbb{D}, \tag{1.17}$$

where W_3 is a continuous positive definite function on $\mathcal{D}$. Then, the origin is uniformly asymptotically stable. Moreover, if $\mathcal{D} = \mathbb{R}^n$, and $W_1(x)$ is radially unbounded, then the origin is globally uniformly asymptotically stable.

Theorem 1.3 (Lyapunov's direct method—Exponential stability theorem). *Let us assume that $x_e = 0$ is an equilibrium point for the system dynamics (1.10), and $\mathcal{D} \subset \mathbb{R}^n$ is a domain containing $x = 0$. Let $V : [0, \infty[\times \mathcal{D} \to \mathbb{R}$ be a continuously differentiable positive definite function such that*

$$k_1 \|x\|^a \le V(t, x) \le k_2 \|x\|^a, \tag{1.18}$$

$$\frac{\partial V}{\partial t} + \frac{\partial V}{\partial x} f(t, x) \le -k_3 \|x\|^a, \quad \forall t \ge 0, \quad \forall x \in \mathbb{D}, \tag{1.19}$$

where k_1, k_2, k_3, and a are positive constants. Then, the origin is exponentially stable. Moreover, if $\mathcal{D} = \mathbb{R}^n$, then the origin is globally asymptotically stable.

Theorem 1.4 (LaSalle-Yoshizawa theorem). *Consider the time-varying system (1.10) and assume $[0, \infty) \times \mathcal{D}$ is a positively invariant set with respect to Eq. (1.10) where $f(t, \cdot)$ is Lipschitz in x, uniformly in t. Assume there exist a $\mathcal{C}^1$ function $V : [0, \infty) \times \mathcal{D} \to \mathbb{R}$, continuous positive definite functions $W_1(\cdot)$ and $W_2(\cdot)$, and a continuous nonnegative function $W(\cdot)$, such that for all $(t, x) \in [0, \infty) \times \mathcal{D}$,*

$$W_1(x) \leq V(t, x) \leq W_2(x),$$
$$\dot{V}(t, x) \leq -W(x) \tag{1.20}$$

hold. Then there exists $\mathcal{D}_0 \subseteq \mathcal{D}$ such that for all $(t_0, x_0) \in [0, \infty) \times \mathcal{D}_0$, $x(t) \to \mathcal{R} \triangleq \{x \in \mathcal{D} : W(x) = 0\}$ as $t \to \infty$. If, in addition, $\mathcal{D} = \mathbb{R}^n$ and $W_1(\cdot)$ is radially unbounded, then for all $(t_0, x_0) \in [0, \infty) \times \mathbb{R}^n$, $x(t) \to \mathcal{R} \triangleq \{x \in \mathbb{R}^n : W(x) = 0\}$ as $t \to \infty$.

The previous theorems are mainly based on the direct Lyapunov approach, where a positive definite function (Lyapunov candidate) is required to prove the stability of a given equilibrium point. Another "easier" method is the so-called Lyapunov indirect method, which relies on the tangent linearization (i.e., Jacobian, of the system's dynamics at the equilibrium point. Some results are recalled next.

Theorem 1.5 (Lyapunov's indirect method—Autonomous systems). *Let $x = 0$ be an equilibrium point of the autonomous nonlinear dynamics $\dot{x} = f(x)$, $f : \mathcal{D} \to \mathbb{R}^n$, where f is $\mathcal{C}^1$, and $\mathcal{D}$ is a neighborhood of the origin. Consider the tangent linearized system, defined by $\dot{x} = Ax$, where $A = \frac{\partial f}{\partial x}|_{x=0}$. Then, the origin is asymptotically stable if $Re(\lambda_i) < 0$, $\forall i = 1, \dots, n$, where $\lambda_i s$ are the eigenvalues of A. Moreover, the origin is unstable if $\exists \lambda_i$, s.t., $Re(\lambda_i) > 0$.*

We can underline here that Theorem 1.5 does not allow us to conclude anything in the case where the tangent linearized system is marginally stable, that is, if some of the eigenvalues of A are in the left-half plan, and some are on the imaginary axes.

A time-varying version of the previous result is given next.

Theorem 1.6 (Lyapunov's indirect method—Nonautonomous systems). *Let $x = 0$ be an equilibrium point of the nonautonomous nonlinear dynamics $\dot{x} = f(t, x)$, $f : [0, \infty[\times \mathcal{D} \to \mathbb{R}^n$, where f is $\mathcal{C}^1$, and $\mathcal{D}$ is a neighborhood of the origin. Consider the tangent linearized system, defined by $\dot{x} = A(t)x$, where $A(t) = \frac{\partial f}{\partial x}(t, x)|_{x=0}$. Assume that A is bounded and Lipschitz on D, uniformly in t. Then, the origin is exponentially stable for the nonlinear dynamics, if it is exponentially stable for the tangent linearized system.*

Another set of definitions which will be useful throughout the book are the notions related to solutions' boundedness and input-to-state stability (ISS); these are defined next.

Definition 1.32 (Uniform boundedness). *The solutions of the dynamical system (1.10) are uniformly bounded if there exists a constant $c > 0$, independent of t_0, and for every $a \in]0, c[$, there exists a constant $\beta(a) > 0$, independent of t_0, such that*

$$\|x(t_0)\| \leq a \Rightarrow \|x(t)\| \leq \beta, \quad \forall t \geq t_0 \geq 0. \tag{1.21}$$

Moreover, the solutions are globally uniformly bounded if Eq. (1.21) holds for any $a \in]0, \infty[$.

Definition 1.33 (Uniform ultimate boundedness). *The solutions of the dynamical system (1.10) are uniformly ultimately bounded with ultimate bound b, if there exist constants $b > 0$, $c > 0$, independent of t_0, and for every $a \in]0, c[$, there is a constant $T(a, b)$, independent of t_0, such that*

$$\|x(t_0)\| \leq a \Rightarrow \|x(t)\| \leq b, \quad \forall t \geq t_0 + T, \ t_0 \geq 0. \tag{1.22}$$

Moreover, the solutions are globally uniformly bounded if Eq. (1.22) holds for any $a \in]0, \infty[$.

Next, let us consider the system

$$\dot{x} = f(t, x, u) \tag{1.23}$$

where $f : [0, \infty) \times \mathbb{R}^n \times \mathbb{R}^{n_a} \to \mathbb{R}^n$ is piecewise continuous in t and locally Lipschitz in x and u, uniformly in t. The input $u(t)$ is piecewise continuous, bounded function of t for all $t \geq 0$.

Definition 1.34. *The system (1.23) is said to be input-to-sate stable (ISS) if there exist a class $\mathcal{KL}$ function β and a class $\mathcal{K}$ function γ such that for any initial state $x(t_0)$ and any bounded input $u(t)$, the solution $x(t)$ exists for all $t \geq t_0$ and satisfies*

$$\|x(t)\| \leq \beta(\|x(t_0)\|, t - t_0) + \gamma(\sup_{t_0 \leq \tau \leq t} \|u(\tau)\|).$$

Theorem 1.7. *Let $V : [0, \infty) \times \mathbb{R}^n \to \mathbb{R}$ be a continuously differentiable function such that*

$$\alpha_1(\|x\|) \leq V(t, x) \leq \alpha_2(\|x\|)$$

$$\frac{\partial V}{\partial t} + \frac{\partial V}{\partial x} f(t, x, u) \leq -W(x), \quad \forall \|x\| \geq \rho(\|u\|) > 0 \tag{1.24}$$

for all $(t, x, u) \in [0, \infty) \times \mathbb{R}^n \times \mathbb{R}^{n_a}$, where α_1, α_2 are class $\mathcal{K}_\infty$ functions, ρ is a class $\mathcal{K}$ function, and $W(x)$ is a continuous positive definite function on $\mathbb{R}^n$. Then, the system (1.23) is ISS.

We also recall the definition of a weaker property than ISS, referred to as local integral ISS.

Definition 1.35. *Consider the system (1.23) where $x \in \mathcal{D} \subseteq \mathbb{R}^n$, $u \in \mathcal{D}_u \subseteq \mathbb{R}^{n_a}$, such that $0 \in \mathcal{D}$, and $f : [0, \infty) \times \mathcal{D} \times \mathcal{D}_u \to \mathbb{R}^n$ is piecewise continuous in t and locally Lipschitz in x and u, uniformly in t. The inputs are assumed to be measurable and locally bounded functions $u : \mathbb{R}_{\geq 0} \to \mathcal{D}_u \subseteq \mathbb{R}^{n_a}$. Given any control $u \in \mathcal{D}_u$ and any $\xi \in \mathcal{D}_0 \subseteq \mathcal{D}$, there is a unique maximal solution of the initial value problem $\dot{x} = f(t, x, u)$, $x(t_0) = \xi$. Without loss of generality, assume $t_0 = 0$. The unique solution is defined on some maximal open interval, and it is denoted by $x(\cdot, \xi, u)$. System (1.23) is locally integral input-to-state stable (LiISS) if there exist functions $\alpha, \gamma \in \mathcal{K}$ and $\beta \in \mathcal{KL}$ such that, for all $\xi \in \mathcal{D}_0$ and all $u \in \mathcal{D}_u$, the solution $x(t, \xi, u)$ is defined for all $t \geq 0$ and*

$$\alpha(\|x(t, \xi, u)\|) \leq \beta(\|\xi\|, t) + \int_0^t \gamma(\|u(s)\|)ds \tag{1.25}$$

for all $t \geq 0$. Equivalently, system (1.23) is LiISS if and only if there exist functions $\beta \in \mathcal{KL}$ and $\gamma_1, \gamma_2 \in \mathcal{K}$ such that

$$\|x(t, \xi, u)\| \leq \beta(\|\xi\|, t) + \gamma_1 \left(\int_0^t \gamma_2(\|u(s)\|)ds \right) \tag{1.26}$$

for all $t \geq 0$, all $\xi \in \mathcal{D}_0$, and all $u \in \mathcal{D}_u$.

Another notion close to ISS is the one relating bounded inputs to bounded outputs, known as input-to-output stability (IOS).

Definition 1.36. *The system (1.23) associated with the output mapping $y = h(t, x, u)$, where $h : [0, \infty) \times \mathbb{R}^n \times \mathbb{R}^{n_a} \to \mathbb{R}^m$ is piecewise continuous in t and continuous in x and u, is said to be IOS if there exist a class $\mathcal{KL}$ function β and a class $\mathcal{K}$ function γ such that for any initial state $x(t_0)$ and any bounded input $u(t)$, the solution $x(t)$ exists for all $t \geq t_0$; that is, y exists for all $t \geq t_0$ and satisfies*

$$\|y(t)\| \leq \beta(\|x(t_0)\|, t - t_0) + \gamma(\sup_{t_0 \leq \tau \leq t} \|u(\tau)\|).$$

Finally, a weaker definition of stability which is based on the boundedness of the solutions is the Lagrange stability.

Definition 1.37. *The system (1.10) is said to be Lagrange stable if for every initial condition x_0 associated with the time instant t_0 there exists $\epsilon(x_0)$, such that $\|x(t)\| < \epsilon, \forall t \geq t_0 \geq 0$.*

We will often deal in this monograph with the specific class of nonlinear models which are affine in the control vector. To this purpose, we will report in the next section some definitions related to this type of model, see Isidori (1989).

1.4 DYNAMICAL SYSTEMS AFFINE IN THE CONTROL

We will present in this section some well-known and useful characteristics of affine nonlinear systems of the form

$$\dot{x} = f(x) + g(x)u, \quad x(0) = x_0,$$
$$y = h(x), \tag{1.27}$$

where $x \in \mathbb{R}^n, u \in \mathbb{R}^{n_a}, y \in \mathbb{R}^m$ $(n_a \geq m)$, represent the state, the input, and the controlled output vectors, respectively, and x_0 is a given finite initial condition. We assume that the vector fields f, columns of g, and function h are such that $f : \mathbb{R}^n \to \mathbb{R}^n$ and the columns of $g : \mathbb{R}^n \to \mathbb{R}^{n \times n_a}$ are smooth vector fields on a bounded set $\mathcal{D}$ of $\mathbb{R}^n$, and $h(x)$ is a smooth function on $\mathcal{D}$.

Definition 1.38. *System (1.27) with $n_a = m$ is said to have a vector relative degree $r = (r_1, \ldots, r_m)^T$ at point x^* if*

1. *$L_{g_j} L_f^k h_i(x) = 0, \ \forall \ 1 \leq j \leq m, \quad \forall \ 1 \leq i \leq m, \quad \forall \ k < r_i - 1$, for all $x \in \mathcal{V}(x^*)$, where $\mathcal{V}(x^*)$ is a neighborhood of x^**
2. *The $m \times m$ matrix*

$$A(x) = \begin{bmatrix} L_{g_1} L_f^{r_1-1} h_1(x) & \cdots & L_{g_m} L_f^{r_1-1} h_1(x) \\ \vdots & \vdots & \vdots \\ L_{g_1} L_f^{r_m-1} h_m(x) & \cdots & L_{g_m} L_f^{r_m-1} h_m(x) \end{bmatrix} \tag{1.28}$$

is none singular at $x = x^*$.

Remark 1.1. *In the general case of rectangular over actuated systems, that is, $n_a > m$, the relative degree vector is defined similar to the square case (with the proper modification of j-index dimension), except that the second condition is replaced by the existence of the pseudoinverse, that is, $A_{\text{left}}^{-1} = (A^T A)^{-1} A^T$.*

1.4.1 Exact Input-Output Linearization by Static-State Feedback and the Notion of Zero Dynamics

Consider the nonlinear system (1.27): the idea of exact input–output linearization is to find a feedback which transforms the nonlinear systems into a linear mapping between a given input vector and an output vector. To do so, we assume that Eq. (1.27) has a vector relative degree $r = (r_1, \ldots, r_m)^T$, s.t., $\tilde{r} \stackrel{\Delta}{=} \sum_{i=1}^{i=m} r_i \leq n$.

Under the previous condition, we can define the diffeomorphism mapping defined in $\mathcal{V}(x^*)$ as

$$\Phi(x) = (\phi_1^1(x), \ldots, \phi_{r_1}^1(x), \ldots, \phi_1^m(x), \ldots, \phi_{r_m}^m(x), \phi_{\tilde{r}+1}(x), \ldots, \phi_n(x))^T, \tag{1.29}$$

with

$$\begin{cases} \phi_1^i = h_i(x), \\ \vdots \\ \phi_{r_i}^i = L_f^{r_i-1} h_i(x), \ 1 \le i \le m. \end{cases} \qquad (1.30)$$

Next, we define the vectors $\xi \in \mathbb{R}^{\tilde{r}}$, and $\eta \in \mathbb{R}^{n-\tilde{r}}$, such that $\Phi(x) = (\xi^T, \eta^T)^T \in \mathbb{R}^n$.

Using these new notations, the system (1.27) writes as

$$\begin{cases} \dot{\xi}_1^i = \xi_2^i, \\ \vdots \\ \dot{\xi}_{r_i-1}^i = \xi_{r_i}^i, \\ \dot{\xi}_{r_i}^i = b_i(\xi, \eta) + \sum_{j=1}^{j=m} a_{ij}(\xi, \eta) u_j, \\ y_i = \xi_1^i, \\ \dot{\eta} = q(\xi, \eta) + p(\xi, \eta) u, \ 1 \le i \le m, \end{cases} \qquad (1.31)$$

where p, q are two vector fields (dependent on f, g, h), with

$$a_{ij} = L_{g_j} L_f^{r_i-1} h_i(\Phi^{-1}(\xi, \eta)), \ 1 \le i, j \le m,$$

and

$$b_i(\xi, \eta) = L_f^{r_i} h_i(\Phi^{-1}(\xi, \eta)), \ 1 \le i \le m.$$

If we define the vectors

$$y^{(r)} \triangleq (y_1^{(r_1)}(t), \ldots, y_m^{(r_m)}(t))^T,$$

and

$$b(\xi, \eta) = (b_1(\xi, \eta), \ldots, b_m(\xi, \eta))^T.$$

Then, the system (1.31) writes as

$$\begin{cases} y^{(r)} = b(\xi, \eta) + A(\xi, \eta) u, \\ \dot{\eta} = q(\xi, \eta) + p(\xi, \eta) u. \end{cases} \qquad (1.32)$$

Next, if we consider a desired output reference vector

$$y_d(t) = (y_{1d}, \ldots, y_{md}(t))^T,$$

and inverse the model (1.32), we obtain the nominal control vector

$$u_d(t) = A^{-1}(\xi_d(t), \eta_d(t))(-b(\xi_d(t), \eta_d(t)) + y_d^{(r)}(t)),$$

where

$$\xi_d(t) = (\xi_d^1(t), \ldots, \xi_d^m(t))^T,$$
$$\xi_d^i(t) = (y_{id}(t), \ldots, y_{id}^{(r_i-1)}(t))^T, \ 1 \leq i \leq m,$$

and $\eta_d(t)$ is a solution of the dynamical system

$$\dot{\eta}_d(t) = q(\xi_d, \eta_d) + p(\xi_d, \eta_d)A^{-1}(\xi_d, \eta_d)(-b(\xi_d, \eta_d) + y_d^{(r)}(t)). \quad (1.33)$$

Definition 1.39. *Eq. (1.33) associated with a nonzero desired trajectory $y_d(t)$ are called internal dynamics of the system (1.27).*

Definition 1.40. *Eq. (1.33) associated with a zero desired trajectory $y_d(t) = 0, \quad \forall t \geq 0$ are called zero dynamics of the system (1.27).*

Assume now, without loss of generality, that the zero dynamics (1.33), where $y_d(t) = 0$ admits $\eta = 0$ as an equilibrium point. Then, we can define the notion of minimum–phase system.

Definition 1.41. *The system (1.27) is minimum phase if the equilibrium point $\eta = 0$ of its zero dynamics is asymptotically stable. The system is nonminimum phase otherwise.*

Definition 1.42. *The system (1.27) is globally minimum phase if the equilibrium point $\eta = 0$ of its zero dynamics is globally asymptotically stable.*

Definition 1.43. *The system (1.27) is weakly minimum phase if there exists a C^r, $r \geq 2$ function W defined on $\mathcal{V}(0)$, such that $W(0) = 0$, $W(x) > 0 \quad \forall x \neq 0$, and the Lie derivative of W along the zero dynamics is nonpositive for all $x \in \mathcal{V}(0)$.*

Definition 1.44. *The system (1.27) is globally weakly minimum phase if there exists a radially unbounded C^r, $r \geq 2$ function W, such that $W(0) = 0$, $W(x) > 0 \quad \forall x \neq 0$, and the Lie derivative of W along the zero dynamics is nonpositive for all x.*

Next, we review a number of basic concepts related to the notions of dissipativity and passivity, see for example, van der Schaft (2000), Brogliato et al. (2007), and Ortega et al. (1998).

Definition 1.45. *A supply rate function for Eq. (1.27) is a scalar function $W(u, y)$ defined on $\mathbb{R}^{na} \times \mathbb{R}^m$, such that $\forall u \in \mathbb{R}^{na}$, $x_0 \in \mathbb{R}^n$, and the output $y(t) = h(x(x_0, t, u))$, we have*

$$\int_0^T |W(u(\tau), y(\tau))|d\tau > \infty, \quad \forall t \geq 0. \quad (1.34)$$

Definition 1.46. *The system (1.27) with a supply rate W is said to be dissipative if a continuous nonnegative function $V : \mathbb{R}^n \to \mathbb{R}$, called a storage function, such that $\forall x_0 \in \mathbb{R}^n$, and $u \in \mathbb{R}^{na}$, we have (the dissipative inequality)*

$$V(x(x_0, t, u)) - V(x_0) \leq \int_0^T W(\tau)d\tau. \quad (1.35)$$

Definition 1.47. *The system* (1.27) *is passive if it is dissipative with supply rate function* $W = u^T y$, *and its storage function* V *satisfies* $V(0) = 0$.

We will end this chapter with some basic geometric and topological definitions, see for example, Alfsen (1971) and Blanchini (1999).

1.5 GEOMETRIC, TOPOLOGICAL, AND INVARIANCE SET PROPERTIES

Definition 1.48. *A metric space is a set* X *with a global distance function* d, *such that, for every two points* $x, y \in X$, *gives the distance between them as a nonnegative real number* $d(x, y)$. *A metric space must also satisfy*

$$d(x, y) = 0, \ \textit{iff } x = y,$$
$$d(x, y) = d(y, x), \qquad\qquad (1.36)$$
$$d(x, y) + d(y, z) \geq d(x, z).$$

Definition 1.49. *A subset* K *of a metric space* X *is compact if, for every collection* $\mathcal{C}$ *of open sets such that* $K \subset \cup_{S \subset \mathcal{C}} S$, *there are finitely many sets* $S_1, \ldots, S_n \in \mathcal{C}$ *such that* $K \subset \cup_{i=1}^{i=n} S_i$. *The collection* $\mathcal{C}$ *is called an open cover for* K *and* $\{S_1, \ldots, S_n\}$ *is a finite subcover for* K.

Definition 1.50. *Given an affine space* E, *a subset* $V \subset E$ *is convex if for any two points* $a, b \in V$, *we have* $c \in V$ *for every point* $c = (1 - \alpha)a + \alpha b$, $\alpha \in [0, 1] \subset \mathbb{R}$.

Examples of convex sets are the empty set, every one-point set $\{a\}$, and the entire affine space E.

Finally, we recall next some notions of set invariance which are often used in control and adaptation. Let us consider the following nonlinear nonautonomous uncertain model

$$\dot{x}(t) = f(x(t), u(t), w(t)), \qquad\qquad (1.37)$$

where $x \in \mathbb{R}^n$ is the state vector, $u \in \mathbb{R}^{na}$ is the control vector, and $w \in \mathcal{W} \subset \mathbb{R}^p$ is the uncertainty vector ($\mathcal{W}$ is a given compact set). We assume that f is Lipschitz, and u, w are at least C^0.

Definition 1.51. *The set* $\mathcal{S} \subset \mathbb{R}^n$ *is said to be positive invariant for the system* (1.37), *without control or uncertainty, that is,* $\dot{x} = f(x)$, *if for all* $x(0) \in \mathcal{S}$ *the solution* $x(t) \in \mathcal{S}$ *for all* $t > 0$.

Definition 1.52. *The set* $\mathcal{S} \subset \mathbb{R}^N$ *is said to be a robustly positively invariant set for the system* (1.37), *without control, that is,* $\dot{x} = f(x, w)$, *if for all* $x(0) \in \mathcal{S}$ *and all* $w \in \mathcal{W}$ *the solution is such that* $x(t) \in \mathcal{S}$ *for all* $t > 0$.

Definition 1.53. *The set* $\mathcal{S} \subset \mathbb{R}^N$ *is said to be a robustly controlled (positively) invariant set for the system* (1.37); *if there exists a continuous feedback* *control law $u(t) = g(x(t))$, which ensures the existence and uniqueness of the solution $x(t) \in \mathbb{R}^n$ defined on $\mathbb{R}^+ \cup \{0\}$, and is such that S is a positive invariant set for the closed-loop system.*

1.6 CONCLUSION

In this chapter, we wanted to present briefly some of the main concepts extensively used in control theory. To keep the presentation succinct, we did not go into too much detail; instead we referred the reader to some textbooks which can be consulted to obtain more details about these concepts. We wanted this chapter to be a quick introduction to such concepts so that the reader can follow the rest of the monograph without the need of extra material. Furthermore, in each of the following chapters, we will start with a brief recap of the definitions needed in the specific chapter.

REFERENCES

Alfsen, E., 1971. Convex Compact Sets and Boundary Integrals. Springer-Verlag, Berlin.

Blanchini, F., 1999. Set invariance in control—a survey. Automatica 35 (11), 1747–1768.

Brogliato, B., Lozano, R., Mashke, B., Egeland, O., 2007. Dissipative Systems Analysis and Control. Springer-Verlag, Great Britain.

Golub, G., Van Loan, C., 1996. Matrix Computations, third ed. The Johns Hopkins University Press, Baltimore, MD.

Isidori, A., 1989. Nonlinear Control Systems, second ed., Communications and Control Engineering Series. Springer-Verlag, Berlin.

Khalil, H., 1996. Nonlinear Systems, second ed. Macmillan, New York.

Liapounoff, M., 1949. Problème Général de la Stabilité du Mouvement. Princeton University Press, Princeton, NJ, Tradui du Russe (M. Liapounoff, 1892, Société mathématique de Kharkow) par M. Édouard Davaux, Ingénieur de la Marine à Toulon.

Lyapunov, A., 1992. The General Problem of the Stability of Motion. Taylor & Francis, Great Britain, with a biography of Lyapunov by V.I. Smirnov and a bibliography of Lyapunov's works by J.F. Barrett.

Ortega, R., Loria, A., Nicklasson, P., Sira-Ramirez, H., 1998. Passivity-Based Control of Euler-Lagrange Systems. Springer-Verlag, Great Britain.

Perko, L., 1996. Differential Equations and Dynamical Systems, Texts in Applied Mathematics. Springer, New York.

Rouche, N., Habets, P., Laloy, M., 1977. Stability Theory by Liapunov's Direct Method, Applied Mathematical Sciences, vol. 22. Springer-Verlag, New York.

van der Schaft, A., 2000. L2-Gain and Passivity Techniques in Nonlinear Control. Springer-Verlag, Great Britain.

Zubov, V., 1964. Methods of A.M. Lyapunov and Their Application. The Pennsylvania State University, State College, PA, translation prepared under the auspices of the United States Atomic Energy Commission.

CHAPTER 2

Adaptive Control: An Overview

2.1 INTRODUCTION

The goal of this chapter is to give the reader an overview of adaptive control in general, and learning-based adaptive control in particular. Indeed, the field of adaptive control is vast and the task of reporting in one venue all the existing results in this area seems pharaonic. Nevertheless, we will try here to report some of the most known results in the field of adaptive control theory. We apologize in advance for any missing results and references. To help us simplify the presentation, we have decided to structure the chapter based on the nature of the adaptive control method. What we mean by nature of the method is whether the adaptive approach is entirely model-based, entirely model-free, or partially model-based. Indeed, because the main idea of this book is based on merging together some model-based controllers and some model-free learning algorithms to design learning-based adaptive controllers, we have decided to review the existing results in adaptive control theory based on their dependence on the system model. We are well aware that this is not the most usual way of classifying results in the control community, where it is more usual to see methodologies classified based on the nature of the model (e.g., linear vs. nonlinear, continuous vs. discrete, etc.). However, we believe that the classification chosen here fits the main purpose of this book.

Before reviewing some results in the field of adaptation and learning, let us first define the terms learn and adapt. Referring to the Oxford dictionary we find these two definitions: Adapt is defined as *to change something in order to make it suitable for a new use or situation, or to change your behavior in order to deal more successfully with a new situation*. As for learn, it is defined as *to gain knowledge and skill by studying, from experience, or to gradually change your attitudes about something so that you behave in a different way*.

From these two definitions we see that both terms refer to changing a given behavior in order to improve the outcome with regard to a given situation. The main difference between the two definitions lies in the fact that *learn* also refers to gaining knowledge from study and experience.

Learning-Based Adaptive Control
http://dx.doi.org/10.1016/B978-0-12-803136-0.00002-6
19

We believe that these definitions also point out the main difference between adaptive control and learning-based adaptive control, which is the fact that learning has to be based on experience. Experience often comes from repeating the same task over and over, while improving the performance at each new iteration based on the mistakes made in previous iterations. We will keep in mind this characterization of learning-based adaptive control throughout the book.

There have been many survey papers on adaptive control theory and applications. Some survey papers were written in the 1970s and 1980s: Jarvis (1975), Astrom (1983), Seborg et al. (1986), Ortega and Tang (1989), Isermann (1982), and Kumar (1985); some papers were written in the 1990s: Egardt (1979), Astrom and Wittenmark (1995), Ljung and Gunnarsson (1990), Filatov and Unbehauen (2000), and Astrom et al. (1994). However, to the best of our knowledge, there have been much fewer survey papers on adaptive control published recently (Tao, 2014; Barkana, 2014). Instead, we found several books on this topic, some with a general presentation of adaptive control, for example, Astrom and Wittenmark (1995), and some others more specialized on a specific subtopic in the adaptive control field, for example, robust adaptive control (Ioannou and Sun, 2012), linear adaptive control (Landau, 1979; Egardt, 1979; Goodwin and Sin, 1984; Narendra and Annaswamy, 1989; Astrom and Wittenmark, 1995; Landau et al., 2011; Sastry and Bodson, 2011; Tao, 2003), nonlinear adaptive control (Krstic et al., 1995; Spooner et al., 2002; Astolfi et al., 2008; Fradkov et al., 1999), stochastic adaptive control (Sragovich, 2006), and learning-based adaptive control (Vrabie et al., 2013; Wang and Hill, 2010; Zhang and Ordóñez, 2012). We will refer to some of these papers and books in this chapter, as well as add newer developments in the field based on more recent publications on adaptive control theory.

Interest in adaptive control algorithms started in the 1950s with the emergence of flight controllers, see Gregory (1959) and Carr (2014). Since then, there has been a great deal of theoretical research and practical applications of adaptation in control systems. On the theoretical side, the early advances were made in systems parametric identification (Astrom and Eykhoff, 1971), dynamics programming stochastic control theory (Bellman, 1957, 1961; Tsypkin, 1971, 1975), where the basic foundations have been formulated for adaptive and learning control theory as we know it today. As for the application side, there have been a large number of applications of adaptive control and learning, either in laboratory test beds or in real-life industrial application; for example, some surveys of applications are given in Seborg et al. (1986), Astrom and Wittenmark (1995), and Carr (2014).

Currently, we can count a large number of papers and results on adaptive and learning control theory. One way to *try to* classify these results is to group them in terms of their correlation to the physical model of the system. Indeed, we can decompose adaptive control theory into the following three main classes: model-based adaptive control, model-free adaptive control, and learning-based adaptive control. To make sure that there is no confusion in the mind of the reader, let us first define what we mean by each class. In this book, when we talk about model-based adaptive controllers, we mean controllers that are *entirely* based on the model of the system which they are designed to control. Next, when we talk about model-free adaptive controllers, we mean controllers that do not rely on any physical model of the system; instead they are fully based on trial and error and learning. Finally, when we talk about learning-based adaptive controllers, we mean controllers that use some basic physical model of the system, albeit partially known, and add to it a learning layer to compensate for the unknown part of the model. The results of this book fall in the later class of adaptive controllers.

Each of these classes can be further decomposed into subclasses. Indeed, when talking about model-based adaptive control, we can identify the following subclasses: direct adaptive control and indirect adaptive control. Direct adaptive control is based on the idea of designing a controller which adapts to the model uncertainties, without explicitly having to identify the true values of these uncertainties. On the other hand, indirect adaptive control aims at estimating the true values of the uncertainties, and then using the uncertainties estimates and their associated estimated model to design the controller. Other subclasses can be identified based on the type of model used to design the controllers, for example, linear (model-based) adaptive controllers, nonlinear (model-based) adaptive controllers, continuous (model-based) adaptive controllers, discrete (model-based) adaptive controllers, and so on. In this chapter, we will try[1] to present the existing approaches following the classifications given previously.

Next, we use the three main classes mentioned previously to classify the available results in three main groups, and then present each group in a

---

[1] The endeavor of reporting in one chapter all the existing work on adaptive control is tedious, not to say impossible, so we do apologize if we missed here some interesting papers in the field.

separate section. We start by presenting the adaptive control problem in the context of general dynamical systems in Section 2.2. Then, in Section 2.3 we review some works related to the model-based adaptive control class. In Section 2.4, we report some results related to the model-free adaptive control class. Finally, in Section 2.5, we report works related to the hybrid[2] class of learning-based adaptive control, which is the main topic of this book. This chapter ends with some concluding remarks in Section 2.6.

2.2 ADAPTIVE CONTROL PROBLEM FORMULATION

The goal of this section is to define the problem of adaptive control in the general setting of nonlinear dynamics. We will keep this formulation as general as possible so that we can use it in different sections of this chapter.

Let us consider a system described in the state space by nonlinear dynamics of the form

$$\begin{aligned} \dot{x} &= f(t, x, u, p(t)), \\ y &= h(t, x, u), \end{aligned} \tag{2.1}$$

where we assume, unless we explicitly say otherwise, that f, h are smooth (up to some required degree); $x \in \mathbb{R}^n$, represents the system states, for example, Eulerian or Lagrangian positions and velocities, and so on; $u \in \mathbb{R}^{n_c}$ represents the control vector, for example, mechanical forces, torques, or electrical currents and voltages; $y \in \mathbb{R}^m$ represents an output vector of interest; and finally $p \in \mathbb{R}^p$ is a vector where we concatenate all the unknown parameters of the model. Note that p can be either constant or time varying. Indeed, in some systems, several constant (physical) parameters are simply unknown to the control designer; in other situations, some parameters might be known originally (their nominal values are known) but they might change over time, for example, drift due to aging of the system, or due to the occurrence of faults in some parts of the system, in which case p will be represented by time-varying variables.

We want also to clarify here that Eq. (2.1) is suitable to define the general problem of parametric adaptive control. However, in some cases, the whole model, that is, f might be unknown, which will be typically the main assumption in model-free adaptive theory. In other cases, part of f is known, but part of it is unknown. For example, consider the case

---

[2]Hybrid in the sense that it uses partly a model-based control, and partly a model-free control.

where $f = f_1(t, x, u) + f_2(t, x, p)$, with f_1 known and f_2 unknown. This case is typically suitable for learning-based adaptive theory, where the control design is based on both the known part of the model f_1, and some learning-based control part to compensate for the unknown part of the model f_2. Other combinations or forms of uncertainties are possible, and will be presented when we report the corresponding adaptive control methods.

2.3 MODEL-BASED ADAPTIVE CONTROL

In this section we want to focus on model-based adaptive control. As we explained earlier, by model based we mean adaptive controllers that are designed completely based on the system's model. We will classify this type of model-based adaptive controller in terms of the approach used to compensate for the model uncertainties, either direct or indirect, but also in terms of the nature of the model and the controller equations. Indeed, we can classify the model-based adaptive algorithm as linear controllers (continuous or discreet) and nonlinear controllers (continuous and discreet). We will present this classification in more detail in the remainder of this section.

2.3.1 Direct Model-Based Adaptive Control

We first want to clarify what we mean by direct adaptive control. We will follow here the literature of adaptive control and define direct adaptive control, as an approach where a feedback adaptive control adjusts *the controller parameters* online to compensate for some model uncertainties. We cite next some of the main references in direct adaptive control by decomposing them into two main groups, namely linear versus nonlinear control.

2.3.1.1 Linear Direct Model-Based Adaptive Control

Linear adaptive control aims at designing adaptive controllers for systems described by linear models, where the state-space model (2.1) in this case writes as

$$\dot{x} = Ax + Bu,$$
$$y = Cx + Du, \tag{2.2}$$

where the matrices A, B, C, and D are of appropriate dimensions, and could be partially known only. The coefficients of these matrices could also be time varying. We want to point out here that by linear models we do not mean exclusively state-space models of the form (2.2). Indeed, a lot of linear adaptive methods have been developed for input-output differential equation (continuous) models, or for input-output difference equation (discreet) models.

There have been a myriad of textbooks and survey papers about model-based linear (direct/indirect) adaptive control. For instance, we can cite the following books: Landau (1979), Egardt (1979), Goodwin and Sin (1984), Narendra and Annaswamy (1989), Astrom and Wittenmark (1995), Ioannou and Sun (2012), Landau et al. (2011), and Sastry and Bodson (2011); and the following survey papers: Isermann (1982), Ortega and Tang (1989), Astrom (1983), Astrom et al. (1994), Seborg et al. (1986), and Barkana (2014). Let us recall here some of the main approaches.

We could further decompose linear adaptive control methods in terms of the main approach used to design the controller. For example, one of the main ideas in direct adaptive control theory relies on Lyapunov stability theory, where the controller and its parameters' adaptation laws are designed to make a given Lyapunov function nonpositive definite. We will refer to direct adaptive controls based on this idea, as direct Lyapunov-based adaptive control methods. Another very well-known approach is the direct model reference adaptive control (d-MRAC), which uses a desired model reference (with constant input reference in case of a regulation problem, or a time-varying reference trajectory in case of a tracking problem), compares its outputs to the system's outputs, and uses the outputs error to adapt the controller parameters in such a way to minimize the outputs error and keep all the feedback signals bounded. It is worth noting here that some d-MRAC methods can be classified as a direct Lyapunov-based adaptive control because they use some specific Lyapunov function (e.g., the strictly positive real (SPR) Lyapunov design approach, Ioannou and Sun (2012)) to design the controller form and its parameters' adaptation laws.

Let us report here a simple example (Ioannou and Sun, 2012), where a Lyapunov-based d-MRAC method is applied to a linear model of the form (2.2) with constant unknown matrices A and B. The reference model is given by

$$\dot{x}_{\mathrm{r}} = A_{\mathrm{r}} x + B_{\mathrm{r}} r, \tag{2.3}$$

where $A_{\mathrm{r}} \in \mathbb{R}^{n \times n}$, $B_{\mathrm{r}} \in \mathbb{R}^{n \times n_{\mathrm{c}}}$ and $r \in \mathbb{R}^{n_{\mathrm{c}}}$ is a bounded smooth (to some desired degree) reference trajectory. We also assume that the pair (A, B) is controllable. The control law is chosen to be of the same form as the reference model

$$u = -K(t)x + L(t)r, \tag{2.4}$$

where K, L are two time-varying matrices, which have to be adapted online to compensate for the unknown model parameters, that is, A, B, and still achieve asymptotic trajectory tracking. Next, we define the ideal control matrices, in such a way that the feedback system matches the reference model, as follows:

$$A - BK^* = A_r, \quad BL^* = B_r. \tag{2.5}$$

We can now define the following errors

$$e = x - x_r,$$
$$K_e = K - K^*, \tag{2.6}$$
$$L_e = L - L^*.$$

After some simple algebraic manipulations of the feedback dynamics, we can write the error dynamics

$$\dot{e} = A_r e + B(-K_e x + L_e r). \tag{2.7}$$

Next, to design the adaptation laws which generate $K(t)$ and $L(t)$, we use the following Lyapunov function

$$V(e, K_e, L_e) = e^T P e + \text{tra}(K_e^T \Gamma K_e + L_e^T \Gamma L_e), \tag{2.8}$$

where tra(.) denotes the trace of a matrix, Γ satisfies $\Gamma^{-1} = L^* \text{sgn}(l)$, with $\text{sgn}(l) = 1$, if $L^* > 0$, $\text{sgn}(l) = -1$, if $L^* < 0$ and $P = P^T > 0$, satisfies the Lyapunov equation

$$PA_e + A_e^T P = -Q, \quad Q = Q^T > 0. \tag{2.9}$$

Notice that the Lyapunov function V is function of all the errors, that is, the tracking error e, but also the control parameters' errors K_e and L_e. Next, a classical Lyapunov approach is used to find an adaptation law for K_e and L_e. Indeed, the computation of the Lyapunov function derivative leads to

$$\dot{V}(e, K_e, L_e) = -e^T Q e + 2e^T P B_r L^{*-1}(-K_e x + L_e r)$$
$$+ 2\text{tra}(K_e^T \Gamma \dot{K}_e + L_e^T \Gamma \dot{L}_e), \tag{2.10}$$

where

$$\text{tra}(x^T K_e^T \Gamma B_r^T P e)\text{sgn}(l) = \text{tra}(K_e^T \Gamma B_r^T P e x^T)\text{sgn}(l) = e^T P B_r L^{*-1} K_e x,$$
$$\text{tra}(L_e^T \Gamma B_r^T P e r^T)\text{sgn}(l) = e^T P B_r L^{*-1} L_e.$$

$$\tag{2.11}$$

Finally, by examining the Lyapunov function derivative (2.10), we can choose the following control parameters' update laws

$$\dot{K}_e = \dot{K} = B_e^T P e x^T \mathrm{sgn}(l),$$
$$\dot{L}_e = \dot{L} = -B_e^T P e r^T \mathrm{sgn}(l). \tag{2.12}$$

This choice leads to the negative derivative

$$\dot{V} = e^T Q e < 0, \tag{2.13}$$

which implies, using Barbalat's Lemma, that K, L, e are bounded, and $e(t) \to 0$, for $t \to \infty$.

This simple example of d-MRAC shows how Lyapunov stability theory is usually used in direct adaptive control to design the control as well as its parameters' update laws. It is worth noting here that other approaches in d-MRAC are based on linear input-output parametrization of the feedback system, and use simple-gradient or least-square algorithms to update the controller parameters (see, e.g., Ioannou and Sun, 2012, Chapter 6).

We are not going to report here more direct Lyapunov-based methods for linear systems, as they all follow a similar procedure to the one presented on this simple case. We refer the reader to the survey papers and books cited previously for more examples and analysis details.

Another well-known method for direct linear adaptive control is direct adaptive pole placement control (d-APPC). This method is conceptually similar to d-MRAC; the difference is that instead of trying to replicate the behavior of a reference model, here one tries to design a desired closed-loop response by placing the closed-loop poles in some desired areas. It is of course motivated by the well-known pole placement approach for linear time-invariant systems, where in this case, the unknown coefficients of the plants are replaced by their estimates (this is known as the certainty equivalence concept). However, in d-APPC, the plants' unknown coefficients are not estimated explicitly; instead, the control coefficients are updated to compensate for the plants' uncertainties.

For completeness, besides the review papers and the specialized textbooks that we mentioned previously, we also want to cite here some more recent works on direct linear adaptive control. One of these works is referred to as l_1 adaptive control for linear systems, see Cao and Hovakimyan (2008). The main idea in l_1 adaptive control can be seen as an extension of d-MRAC, where a strictly proper stable filter is inserted before the d-MRAC controller to guarantee some transient tracking performances (providing high adaptation feedback), on top of the asymptotic performances obtained by classical d–MRAC. However, the claimed performances of this approach have been challenged in Ioannou et al. (2014).

We can also cite works that focus on designing direct adaptive controllers for linear systems under input constraints, see Park et al. (2012), Yeh and Kokotovit (1995), Rojas et al. (2002), and Ajami (2005). Finally, we could classify as direct linear adaptive control the concept of multiple-models adaptive switching control, where a set of robust linear controllers is designed to deal with a range of plants' uncertainties, and a supervisor is used to switch online between these controllers, while maintaining performance and stability, see Narendra and Balakrishnan (1997), Narendra et al. (2003), Giovanini et al. (2005), and Giovanini et al. (2014).

2.3.1.2 Nonlinear Direct Model-Based Adaptive Control

Some of the concepts mentioned previously have been extended to the more challenging case of nonlinear systems. Following an early formulation, which appeared in the Russian literature in the late 1960s, see Yakubovich (1969) and Fradkov (1994), we can state the model-based nonlinear adaptive control problem as follows: consider a nonlinear plant modeled in the state space by the nonlinear differential equation

$$\dot{x} = F(t, x, u, p),$$
$$y = G(t, x, u, p), \tag{2.14}$$

where $t \in \mathbb{R}^+$, $x \in \mathbb{R}^n$, $y \in \mathbb{R}^m$, and $u \in \mathbb{R}^{n_c}$ are the scalar time variable, the state vector, the output vector, and the control vector, respectively. $p \in \mathcal{P} \subset \mathbb{R}^p$ represents a vector of unknown parameters, element of an a priori known set $\mathcal{P}$. F and G are two smooth functions. Next, we define a control objective function $Q(t, x, u) : \mathbb{R}^+ \times \mathbb{R}^n \times \mathbb{R}^{n_c} \rightarrow \mathbb{R}$, such that the control objective is deemed reached iff

$$Q(t, x, u) \leq 0, \quad \text{for } t \geq t_* > 0. \tag{2.15}$$

The adaptive problem is to find a (model-based) two-level control algorithm, of the form

$$u(t) = U_t(u(s), y(s), \theta(s)), \quad s \in [0, t[,$$
$$\theta(t) = \Theta_t(u(s), y(s), \theta(s)), \quad s \in [0, t[, \tag{2.16}$$

where U_t and Θ_t are nonanticipative operators (or causal operators). The goal of the two-level control (2.16) is to meet the control objective (2.15), while maintaining the closed-loop trajectories of Eqs. (2.14), (2.16) bounded for any $p \in \mathcal{P}$, and for a given initial conditions set Ω, s.t., $(x(0), \theta(0)) \in \Omega$. This formulation is indeed fairly general. For instance, if one is concerned with a regulation adaptive problem, the objective function Q can be chosen as:

- For state regulation to a constant vector x_*:

$$Q_{xr} = \|x - x_*\|_R^2 - \delta, \quad R = R^T > 0, \ \delta > 0. \quad (2.17)$$

- For output regulation to a constant vector y_*:

$$Q_{yr} = \|y - y_*\|_R^2 - \delta, \quad R = R^T > 0, \ \delta > 0. \quad (2.18)$$

Similarly, if state or output tracking is the control target, then the objective function can be formulated as

$$Q_{xt} = \|x - x_*(t)\|_R^2 - \delta, \quad R = R^T > 0, \ \text{(state tracking of } x_*(t)),$$
$$Q_{yt} = \|y - y_*(t)\|_R^2 - \delta, \quad R = R^T > 0, \ \text{(output tracking of } y_*(t)),$$
$$(2.19)$$

where the reference trajectories are solutions of a desired reference model

$$\dot{x}_* = F_*(t, x_*),$$
$$y_* = G_*(t, x_*). \quad (2.20)$$

Of course, to be able to obtain any useful (and maybe rigorous) analysis of the nonlinear controllers, some assumptions on the type of nonlinearities had to be formulated. For instance, a very well-known assumption in nonlinear (both direct and indirect) adaptive control is the linear parametrization of the plant's model, that is, F and G in Eq. (2.14), as a function of the plant's unknown parameters p. We think that this assumption is important enough to be used as a way to further classify direct nonlinear controllers, in terms of the nature of the plant's parametrization in the unknown parameters. In the following, we will classify some of the available results in direct nonlinear adaptive control in two main subclasses: linear parametrization-based controllers versus nonlinear parametrization-based controllers.

Let us first cite some of the available monographs and survey papers on model-based nonlinear adaptive control (of course including both direct and indirect approaches). We can cite the following books: Slotine and Li (1991, Chapter 8), Krstic et al. (1995), Spooner et al. (2002), and Astolfi et al. (2008); and the following survey papers: Ortega and Tang (1989), Ilchmann and Ryan (2008), Barkana (2014), and Fekri et al. (2006).

We start by recalling, on a simple example, a direct Lyapunov-based approach applied in the case of a linear parametrization, for example, for more details the reader can refer to Krstic et al. (1995, Chapter 3). Let us consider the simple nonlinear dynamical system

$$\dot{x} = \theta \phi(x) + u, \tag{2.21}$$

where for simplicity, we assume that $\theta \in \mathbb{R}$ is a constant unknown parameter, $x \in \mathbb{R}$, $u \in \mathbb{R}$, and $\phi(.)$ is a nonlinear function of x.

In this case a straightforward controller can be written as

$$u = -\theta \phi(x) - cx, \quad c > 0, \tag{2.22}$$

which would be an asymptotically stabilizing controller if θ was known. However, we assumed that θ was unknown, so instead of its actual value we use its estimated value $\hat{\theta}$ (certainty equivalence principle). In that case, the control will write as

$$u = -\hat{\theta} \phi(x) - cx, \quad c > 0, \tag{2.23}$$

where $\hat{\theta}$ needs to be updated online following some adaptive update law. To find a suitable update law, we can use a Lyapunov-based approach as follows: First, let us consider the following Lyapunov function candidate

$$V = \frac{1}{2}(x^2 + \theta_e^2), \tag{2.24}$$

where $\theta_e = \theta - \hat{\theta}$ is the parameters' estimation error. Next, the derivative of V along the closed-loop dynamics leads to

$$\dot{V} = -cx^2 + \theta_e(\dot{\theta}_e + x\phi(x)). \tag{2.25}$$

By examining the expression of $\dot{V}$ we can choose the following update law for $\hat{\theta}$

$$\dot{\hat{\theta}} = -\dot{\theta}_e = x\phi(x). \tag{2.26}$$

This choice of update law leads to

$$\dot{V} = -cx^2, \tag{2.27}$$

which implies boundedness of x, θ_e, as well as convergence of x to zero (by LaSalle-Yoshizawa theorem). Note that this adaptive controller does not lead to the estimation of the true value of θ. So in this case, $\hat{\theta}$ can be interpreted as a simple feedback control parameter, which is tuned online based on the adaptation law (2.26). Thus, the controller (2.23), (2.26) can be classified as a direct adaptive approach.

As mentioned previously, another important problem in adaptive control is the case where the unknown parameters do not appear in the plant's model as a linear function. This case is usually referred to as nonlinear parametrization of the uncertain system. There have been a lot of efforts to deal with such cases of nonlinear parametrization. For instance we could cite the following papers: Seron et al. (1995), Ortega and Fradkov (1993), Fradkov (1980), Karsenti et al. (1996), Fradkov and Stotsky (1992), Fradkov et al. (2001), Liu et al. (2010), Flores-Perez et al. (2013), and Hung et al. (2008). Let us present on a simple example one of these results, namely the speed gradient-based adaptive method for nonlinear parametrization, which was introduced first by the Russian literature in Krasovskii (1976), Krasovskii and Shendrik (1977), and Fradkov et al. (1999).

To begin with, we present the general principle of the speed-gradient control methods, see Fradkov (1980). Consider a general nonlinear "adjustable" dynamic of the form

$$\dot{x} = F(t, x, p), \tag{2.28}$$

where t, x, and F are the usual time, state variables, and a function (smooth and nonsingular) modeling the studied dynamics. The important object in Eq. (2.28) is the vector p of adjustable variables, which can be considered as control actions, or in our context as the adjustable parameters θ in a closed-loop adaptive system. Then, the speed-gradient approach is based on the minimization of a given smooth (nonsingular) control objective function Q. The speed-gradient method regulates the adjustable vector p in the direction of the gradient of the rate of change of Q, that is, gradient of Q. In this case, a required assumption is that $\dot{Q}$ is convex in p. For instance we consider $Q(t, x) \geq 0$, and the objective is to asymptotically drive Q to zero. Then, if we define $\omega(t, x, p) = \dot{Q} = \frac{\partial Q}{\partial t} + (\nabla_x Q)^T F(t, x, p)$, the speed-gradient update law

$$\dot{p} = -\Gamma \nabla_p \omega(t, x, p), \quad \Gamma = \Gamma^T > 0, \tag{2.29}$$

under the condition of convexity of ω w.r.t. p, and the so-called asymptotic stabilizability condition, see Fradkov (1994, p. 427): $\exists p^*$ s.t. $\omega(t, p^*, x) \leq -\rho(Q)$, $\rho(Q) > 0 \; \forall Q > 0$, the control objective function Q tends to zero asymptotically. This chapter is not about the speed-gradient methods per se, so we will not go here into more detail about these types of control approaches. Instead, let us go back to the focus of this book, that is, adaptive control, and present one of the applications of speed gradient to adaptive control.

We will report next an application of speed gradient to passivity-based adaptive control, introduced in Seron et al. (1995). The goal here is to use passivity characteristics to solve nonlinear adaptive control problems of the general form (2.14), (2.15), and (2.16). The control objective is then to render an uncertain nonlinear system passive (also known as passifying the system), using an adaptive controller of the form (2.16). For instance, based on the definition of input-output passivity (please refer to Chapter 1), one can choose the following control objective function (2.15)

$$Q^p = V(x(t)) - V(x(0)) - \int_0^t u^T(\tau)y(\tau)d\tau, \tag{2.30}$$

where V is a C^r real-valued nonnegative function, s.t., $V(0) = 0$. The reader can see that this choice of control objective function is due to the definition of input-output C^r passivity. Indeed, if the control objective (2.15) is (adaptively) achieved, then one can write the inequality

$$\int_0^t u^T(\tau)y(\tau)d\tau \geq V(x(t)) - V(x(0)), \tag{2.31}$$

which implies passivity of the system. Because the stability properties of passive systems are well documented, see Byrnes et al. (1991), one can then easily claim asymptotic stability of the adaptive feedback system. For example, let us consider the following nonlinear system in the following normal form

$$\begin{aligned}
\dot{z} &= f_1(z, p) + f_2(z, y, p)y, \\
\dot{y} &= f(z, y, p) + g(z, y)u,
\end{aligned} \tag{2.32}$$

where $z \in \mathbb{R}^{n-m}$ are the zero-dynamics' states, $y \in \mathbb{R}^m$, and $p \in \mathbb{R}^p \in \mathcal{P}$ is the vector of constant unknown parameters. We assume that g is nonsingular. We also assume that the system (2.32) is weakly minimum-phase (please refer to Chapter 1 for the definition of minimum-phase systems). Then, one can propose the following nonlinear (passifying) adaptive controller

$$\begin{aligned}
u(z, y, \theta) &= g^{-1}(z, y)(w^T(z, y, \theta) - \rho y + v), \quad \rho > 0, \\
\dot{\theta} &= -\Gamma \nabla_\theta(w(z, y, \theta)y), \quad \Gamma > 0.
\end{aligned} \tag{2.33}$$

Where w is defined as

$$w(z, y, \theta) = -f^T(z, y, \theta) - \frac{\partial W(z, \theta)}{\partial z} f_2(z, y, \theta), \tag{2.34}$$

with W satisfies the following (the weak minimum–phase) condition

$$\frac{\partial W(z,\theta)}{\partial z} f_1(z,\theta) \leq 0, \qquad (2.35)$$

locally in z, $\forall \theta \in \mathcal{P}$. Under the assumption of convexity of the function $w(z,y,\theta)y$ w.r.t. $\theta \in \mathcal{P}$, the adaptive closed-loop (2.32), (2.33), and (2.34), is passive from the new input v to the output y. Finally, asymptotic stabilization can be easily achieved by a simple output feedback from y to v.

This type of direct adaptive control has been used to solve many real-life adaptive problems. For example, in Benosman and Atinc (2013b) a nonlinear direct backstepping-based adaptive controller has been proposed for electromagnetic actuators, and in Rusnak et al. (2014) a missile autopilot based on direct linear adaptive control has been proposed.

Let us now consider the second type of adaptive controllers, namely, indirect adaptive controllers.

2.3.2 Indirect Model-Based Adaptive Control

Similar to the direct adaptive control case, we start by defining what we mean by indirect adaptive control. Following the literature of adaptive control, we define indirect adaptive methods for plants with unknown parameters as control schemes which use a feedback control law complemented with an adaptation law. The goal of the adaptation law here is *to estimate online the true values of the unknown plant's parameters.*

Here again, to simplify the presentation, we classify the available indirect adaptive methods as linear and nonlinear methods.

2.3.2.1 Linear Indirect Model-Based Adaptive Control

We cited in the previous section the existing (to the best of our knowledge) surveys and monographs covering the subject of linear adaptive control. In this section, we choose to illustrate the indirect approach philosophy by presenting on a simple example one of the indirect linear adaptive control methods, namely, indirect model reference adaptive control (ind-MRAC), see Ioannou and Sun (2012, Chapter 6).

Let us consider a simple uncertain model of the form

$$\dot{x} = ax + bu, \qquad (2.36)$$

where all the variables are considered to be scalar, and a and b are two unknown constants. We also assume (to simplify the example) that we know the lower bound $b_{\min}$ of $|b|$. The goal is to design an indirect adaptive controller which achieves tracking of the reference model

$$\dot{x}_m = -a_m x_m + b_m r, \qquad (2.37)$$

where $a_m > 0$, b_m, and $r(t)$ are selected such that the solution of Eq. (2.37), that is, $x_m(t)$ has a desired state profile. We follow here a Lyapunov-based design of the update law to estimate a, b (other designs can be obtained using gradient-descent or least-squares formulation; see, e.g., Ioannou and Sun (2012, Chapter 4)). To do so, we use the following simple control law (which mimics the reference model expression)

$$u = -\frac{a_m + \hat{a}}{\hat{b}} x + \frac{b_m}{\hat{b}} r, \qquad (2.38)$$

where $\hat{a}$ and $\hat{b}$ are the estimates of a and b. To design the update laws of $\hat{a}$ and $\hat{b}$, we use the following Lyapunov candidate

$$V = \frac{1}{2} e_x^2 + \frac{1}{2\gamma_1} a_e^2 + \frac{1}{2\gamma_2} b_e^2, \quad \gamma_1, \gamma_2 > 0, \qquad (2.39)$$

where $e_x = x - x_m$, $a_e = \hat{a} - a$, and $b_e = \hat{b} - b$ are the state tracking error and the plants parameters' estimation errors, respectively. These errors satisfy the error dynamic, obtained from simple substraction of the actual plant from the reference model

$$\dot{e}_x = -a_m e_x - e_a x - e_b u. \qquad (2.40)$$

We can see that the Lyapunov function is similar to the one used in the direct adaptive control cases (2.8), (2.24), in the sense that they all have positive terms of the states' (or outputs') tracking errors, as well as positive terms of the parameters' update errors. Another important point in this example is the possible singularity of the controller (2.38), at $\hat{b} = 0$. To avoid this singularity problem, one straightforward "trick" is to bound the evolution of $\hat{b}$ away from the singularity by using what is known as projection-based adaptation laws, see Ioannou and Sun (2012, Chapter 4).

In this case, the adaptation law for $\hat{a}$ and $\hat{b}$ writes as

$$\dot{\hat{a}} = \gamma_1 e_1 x, \qquad (2.41)$$

$$\dot{\hat{b}} = \begin{cases} \gamma_2 e_1 u \text{ if } |\hat{b}| > b_0, \\ \gamma_2 e_1 u \text{ if } |\hat{b}| = b_0 \text{ and } e_1 u \, \mathrm{sgn}(\hat{b}) \geq 0, \\ 0 \text{ otherwise,} \end{cases} \qquad (2.42)$$

with $|\hat{b}(0)| \geq b_0$. Next, following a classical Lyapunov analysis, we compute the derivative of V along the closed-loop dynamics, and finally one can write

$$\dot{V} \leq -a_m e_1^2 \leq 0, \quad \forall t \geq 0. \tag{2.43}$$

This last inequality allows us to conclude about the boundedness of e_1, $\hat{a}$, and $\hat{b}$, with (using LaSalle-Yoshizawa Theorem) the asymptotic convergence of e_1 to zero. Next, using Barbalat's Lemma one can conclude about the asymptotic convergence of $\dot{\hat{a}}$ and $\dot{\hat{b}}$ to zero. Finally, using a persistence of excitation-type condition (e.g., Ioannou and Sun, 2012, Corollary 4.3.2) on the reference signal r, one can conclude about the exponential convergence of $\hat{a}$ and $\hat{b}$ to a and b, respectively.

This simple example shows us how a Lyapunov-based approach can be used to design and analyze an indirect model-based adaptive controller for linear systems. Let us now consider the more challenging case of nonlinear systems.

2.3.2.2 Nonlinear Indirect Model-Based Adaptive Control

We have cited in the nonlinear direct adaptive control section the available textbooks and survey papers in nonlinear adaptive control. Here again, we will use examples to show some of the general ideas used in nonlinear indirect adaptive control.

First, we want to talk about a very important approach in nonlinear indirect adaptation. This approach, known as *modular design*, has been presented in detail in Krstic et al. (1995, Chapters 5 and 6). The main idea in the modular design is to decompose the control design into two phases. In the first phase, a (nonlinear) controller is designed, which ensures some type of input-to-state boundedness regardless of the model uncertainties' values, for example, input-to-state stability (ISS) (please refer to Chapter 1 for the definition of ISS). Next, in the second phase, some model-based filters are designed to estimate the true values of the model uncertainties. The combination of the two phases leads to an indirect adaptive approach. This approach departs from other certainty equivalence-based adaptive control in the sense that the boundedness of the closed-loop signals is ensured (by the ISS property) for any bounded estimation error of the plant's uncertainties.

We will present next a simple example of this modular design, based on the so-called ISS backstepping (for the first phase), and gradient descent filters (for the second phase).

We consider the following state-space model

$$\dot{x}_1 = x_2 + \phi(x_1)\theta,$$
$$\dot{x}_2 = u, \qquad (2.44)$$

where x_i, $i = 1, 2$ are the states' variables, u the control variable, and $\phi(.)$ a nonlinear scalar function. In the first phase, we want to design a controller which ensures ISS between a given input-output pair. To do so, we use the ISS-backstepping approach. First, we assume that $x_2 = v$ is a new virtual input, and then we find a control feedback which stabilizes the first equation of Eq. (2.44). We use the following $V_1 = \frac{1}{2}x_1^2$. The derivative of V_1 along the first equation of Eq. (2.44), with $v = x_2$ leads to

$$\dot{V}_1 = x_1(v + \phi(x_1)\theta), \qquad (2.45)$$

to ensure ISS between the estimation error of θ, and x_1 we can use the virtual control law

$$v = -c_1 x_1 - \phi(x_1)\hat{\theta} - c_2|\phi(x_1)|^2 x_1, \quad c_1, \ c_2 > 0, \qquad (2.46)$$

with $\hat{\theta}$ being the estimate of θ. In this case, the derivative $\dot{V}_1$ writes as

$$\dot{V}_1 = -c_1 x_1^2 + x_1 \phi(x_1)e_\theta - c_2|\phi|^2 x_1^2, \qquad (2.47)$$

where $e_\theta = \theta - \hat{\theta}$. The right-hand side of Eq. (2.47) can be manipulated a bit to lead to

$$\dot{V}_1 = -c_1 x_1^2 + x_1 \phi(x_1)e_\theta - c_2|\phi|^2 x_1^2 + \frac{1}{4c_2}|e_\theta|^2 - \frac{1}{4c_2}|e_\theta|^2, \qquad (2.48)$$

and by manipulating the squares, one can write

$$\dot{V}_1 = -c_1 x_1^2 - \left(\sqrt{c_2}x_1\phi(x_1) - \frac{1}{2\sqrt{c_2}}e_\theta\right)^2 + \frac{1}{4c_2}|e_\theta|^2, \qquad (2.49)$$

which finally leads to

$$\dot{V}_1 \leq -c_1 x_1^2 + \frac{1}{4c_2}|e_\theta|^2, \qquad (2.50)$$

which implies ISS between x_1 and e_θ (please refer to Chapter 1 for the definition of an ISS Lyapunov function). Next, we go back to the full model (2.44), and design the actual control u which takes into account the error between the virtual control v and x_2. To do so, we use the following (augmented) Lyapunov function

$$V_2 = V_1 + \frac{1}{2}(x_2 - v)^2, \tag{2.51}$$

where v is given by Eq. (2.46). To simplify the notations we denote $z_1 = x_1$, $z_2 = x_2 - v$. The derivative of V_2 along the full model (2.44) leads to

$$
\begin{aligned}
\dot{V}_2 &\leq -c_1 z_1^2 + \frac{1}{4c_2}|e_\theta|^2 + z_1 z_2 + z_2 \left(u - \frac{\partial v}{\partial z_1}(z_2 + v + \phi\theta) - \frac{\partial v}{\partial \hat\theta}\dot{\hat\theta} \right), \\
&\leq -c_1 z_1^2 + \frac{1}{4c_2}|e_\theta|^2 + z_2 \left(u + z_1 - \frac{\partial v}{\partial z_1}(z_2 + v + \phi\hat\theta) \right) \\
&\quad - z_2 \left(\frac{\partial v}{\partial z_1}\phi e_\theta + \frac{\partial v}{\partial \hat\theta}\dot{\hat\theta} \right).
\end{aligned}
\tag{2.52}
$$

Finally, to be able to make V_2 an ISS Lyapunov function for the system with states z_1, z_2, and inputs e_θ, $\hat\theta$ we can choose u in such a way to create square terms, as we did in Eqs. (2.48), (2.49). We consider the following control

$$u = -z_1 - c_3 z_2 - c_4 \left| \frac{\partial v}{\partial z_1}\phi \right|^2 - c_5 \left| \frac{\partial v}{\partial \hat\theta} \right|^2 z_2 + \frac{\partial v}{\partial z_1}(z_2 + v + \phi\hat\theta), \tag{2.53}$$

with c_3, c_4, and c_5 strictly positive coefficients. Substituting u in Eq. (2.52), and performing some algebraic completion of squares, leads to

$$\dot{V}_2 \leq -c_1 z_1^2 - c_3 z_2^2 + \left(\frac{1}{4c_2} + \frac{1}{4c_4} \right)|e_\theta|^2 + \frac{1}{4c_5}|\dot{\hat\theta}|^2, \tag{2.54}$$

which proves the ISS between the states z_1, z_2 and the inputs e_θ, $\dot{\hat\theta}$. Based on the definition of ISS (refer to Chapter 1), we can then write that

$$|z| \leq \beta(z(0), t) + \gamma \sup_{0 \leq \tau \leq t} |\Theta(\tau)|, \quad \forall t \geq 0, \ \gamma > 0, \tag{2.55}$$

where $z = (z_1, z_2)^T$, $\Theta = (e_\theta, \dot{\hat\theta})^T$, and β is a class $\mathcal{KL}$ function. Now that we have ensured that the tracking errors z_1 and z_2 are upper-bounded by the estimation errors e_θ and $\dot{\hat\theta}$, we will use some model-based estimation filters to drive asymptotically the estimation errors to zero. If this is achieved, then by the ISS characteristic, we also drive the tracking error asymptotically to zero. One way to design the estimation filters is to use a gradient-based filter, see Krstic et al. (1995, Chapter 6). For instance, θ can be estimated using the gradient-based filter

$$
\begin{aligned}
\dot{\Omega} &= (A_0 - \lambda \phi^2 P)\Omega + \phi, \\
\dot{\Omega}_0 &= (A_0 - \lambda \phi^2 P)(\Omega_0 + x_1) - x_2, \\
\epsilon &= x_1 + \Omega_0 - \Omega \hat{\theta}, \\
\dot{\hat{\theta}} &= \Gamma \frac{\Omega \epsilon}{1 + \mu |\Omega|^2}, \quad \Gamma > 0, \ \mu > 0
\end{aligned}
\tag{2.56}
$$

where $A_0, P > 0$ are such that $PA_0 + A_0^T P = -I$ (or its rather trivial scalar version in our example). These gradient-based filters are proven to asymptotically converge to the true parameters' values, see Krstic et al. (1995, Lemma 6.5).

We saw in this simple example the logic behind the modular indirect adaptive control for nonlinear systems. This approach can be used to solve more complicated real-life systems. For instance, we refer the reader to Benosman and Atinc (2015), where a modular approach has been proposed, using ISS-backstepping together with gradient-based estimation filters, to solve an adaptive trajectory tracking problem for an electromagnetic actuator.

We also want to report a work on a special case of nonlinear adaptive control, namely, adaptive control for nonlinear nonminimum phase systems (please refer to Chapter 1 for the definition of nonminimum phase systems). We report the results proposed in Liao et al. (2009, 2010), where the authors studied the problem of adaptive control for a general class of nonlinear systems with instable internal dynamics. The authors used a model reference control and a control allocation technique to control the nonlinear system subject to control constraints and internal dynamics stabilization. A Lyapunov design approach was used to estimate some unknown parameters of the system, and ensure that the closed-loop nonlinear system is input-to-state stable.

Finally, to end this section about model-based adaptive control, we want to cite the so-called *composite (or combined) adaptive control*, see Lavretsky (2009), Duarte and Narendra (1989), and Slotine and Li (1989). In this approach, the idea is to combine both the advantages of direct and indirect adaptive control, and write a combined adaptive controller which has been claimed to give better transient behavior than pure direct or indirect MRAC adaptive controllers. However, because these controllers' goal is (partly) estimating the true parameters of the controlled plant, based on our definition of indirect adaptive control, we will classify them in this book among the indirect adaptive controllers.

2.4 MODEL-FREE ADAPTIVE CONTROL

As we mentioned earlier in this chapter, here we mean by model-free adaptive control all controllers that do not rely on any mathematical model of the system. These controllers are solely based on online measurements collected directly from the system. The term adaptive here means that the controller can adapt and deal with any uncertainty in the system because it does not rely on any specific model. For example, one well-known approach which can be used in a model-free control framework is the so-called extremum-seeking (ES) methods, see Zhang and Ordóñez (2012) and Ariyur and Krstic (2003). These types of model-free optimization methods have been proposed in the French literature related to train systems in the 1920s, see Leblanc (1922). Their basic goal is to seek an extremum, that is, to maximize (or minimize) a given function without closed-form knowledge of the function or its gradient. There have been a lot of results on ES algorithms, especially after the appearance of a rigorous convergence analysis in Krstić and Wang (2000). For instance, we can cite the following papers: Ariyur and Krstic (2003), Krstic (2000), Ariyur and Krstic (2002), Tan et al. (2006), Nesic (2009), Tan et al. (2008), Rotea (2000), Guay et al. (2013), Jones et al. (1993), Scheinker and Krstic (2013), Scheinker (2013), Khong et al. (2013b), Noase et al. (2011), Ye and Hu (2013), Tan et al. (2013), Liu and Krstic (2014), and Liu and Krstic (2015).

To give the reader a sense of how ES methods work, let us present some simple ES algorithms. Consider the following general dynamics

$$\dot{x} = f(x, u), \tag{2.57}$$

where $x \in \mathbb{R}^n$ is the state, $u \in \mathbb{R}$ is the scalar control (for simplicity), and $f : \mathbb{R}^n \times \mathbb{R} \to \mathbb{R}^n$ is a smooth function. Now, imagine that Eq. (2.57) represents the model of a real system, and that the goal of the control is to optimize a given performance of the system. This performance can be as simple as a regulation of a given output of the system to a desired constant value, or a more involved output tracking of a desired time-varying trajectory, and so on. Let us now model this desired performance as a smooth function $J(x, u) : \mathbb{R}^n \times \mathbb{R} \to \mathbb{R}$, which we simply denote $J(u)$ because the state vector x is driven by u. To be able to write some convergence results we need the following assumptions.

Assumption 2.1. *There exists a smooth function* $l : \mathbb{R} \to \mathbb{R}^n$ *such that*

$$f(x, u) = 0, \quad \text{if and only if} \quad x = l(u). \tag{2.58}$$

Assumption 2.2. *For each* $u \in \mathbb{R}$ *the equilibrium* $x = l(u)$ *of the system* (2.57) *is locally exponentially stable.*

Assumption 2.3. *There exists (a maximum)* $u^* \in \mathbb{R}$, *such that*

$$(J \circ l)^{(1)}(u^*) = 0,$$

$$(J \circ l)^{(2)}(u^*) < 0. \tag{2.59}$$

Then, based on these assumptions, one can design some simple extremum seekers with proven convergence bounds. Indeed, one of the simplest ways to maximize J is to use a gradient-based ES control as follows:

$$\dot{u} = k \frac{dJ}{du}, \quad k > 0. \tag{2.60}$$

We can analyze the convergence of the ES algorithm (2.60) by using the Lyapunov function

$$V = J(u^*) - J(u) > 0, \quad \text{for } u \neq u^*. \tag{2.61}$$

The derivative of V leads to

$$\dot{V} = \frac{dJ}{du} \dot{u} = -k \left(\frac{dJ}{du} \right)^2 \leq 0. \tag{2.62}$$

This proves that the algorithm (2.60) drives u to the invariant set s.t. $\frac{dJ}{du} = 0$, which is (by Assumption 2.3) equivalent to $u = u^*$. However, as simple as algorithm (2.60) might seem, it still requires the knowledge of the gradient of J. To overcome this requirement, one can use instead an algorithm motivated by the sliding-mode control ideas. For instance, we can define the tracking error

$$e = J(u) - ref(t), \tag{2.63}$$

where *ref* denotes a time function which is monotonically increasing. The idea is that if J tracks *ref*, then it will increase until it reaches an invariant set centered around the equality $\frac{dJ}{du} = 0$. A simple way to achieve this goal is by choosing the following ES law:

$$\dot{u} = k_1 \text{sgn} \left(\sin \left(\frac{\pi e}{k_2} \right) \right), \quad k_1, k_2 > 0. \tag{2.64}$$

This controller is shown to steer u to the set s.t. $|\frac{dJ}{du}| < |\dot{ref}(t)|/k_1$, which can be made arbitrarily small by the proper tuning of k_1, see Drakunov and Ozguner (1992).

Another well-known ES approach is the so-called perturbation-based ES. It uses a perturbation signal (often sinusoidal) to explore the space of control and steers the control variable toward its local optimum by implicitly following a gradient update. These types of ES algorithms have been thoroughly analyzed, for example, in Krstić and Wang (2000), Ariyur and Krstic (2003), Tan et al. (2008), and Rotea (2000). Let us show a simplified version of a sinusoidal disturbance-based ES algorithm.

$$\dot{z} = a\sin\left(\omega t + \frac{\pi}{2}\right)J(u),$$
$$u = z + a\sin\left(\omega t - \frac{\pi}{2}\right), \quad a > 0, \; \omega > 0. \tag{2.65}$$

It has been shown, using averaging theory and singular perturbation theory, that this simple algorithm under some simple assumptions (of at least local optimality and smoothness of J) can (locally) converge to a neighborhood of the optimal control u^*, see Krstić and Wang (2000) and Rotea (2000). There are of course many other ES algorithms; however, it is not the purpose of this chapter to review all the ES results. Instead, we refer the interested reader to the ES literature cited above for more details.

Let us now talk about another well-known model-free control method, namely, the reinforcement learning algorithms, see Busonio et al. (2008), Sutton and Barto (1998), Bertsekas and Tsitsiklis (1996), Szepesvári (2010), Busoniu et al. (2010), Farahmand (2011), and Kormushev et al. (2010). The idea behind reinforcement learning is that by trying random control actions, the controller can eventually build a predictive model of the system on which it is operating. Reinforcement learning is a class of machine learning algorithms which learns how to map states to actions in such a way to maximize a desired reward. In these algorithms the controller has to discover the best actions by trial and error. This idea was motivated by the field of psychology, where it has been realized that animals have the tendency to reselect (or not) actions based on their good (or bad) outcomes, see Thorndike (1911). In reinforcement learning the controller learns an optimal policy (or action), which defines the system's way of behaving at a given time and state. The obtention of the best policy is based on the optimization, through trial and error, of a desired value function. The value function evaluates the value of a policy in the long run. Simply put, the value function at a given state is the total amount of immediate reward a controller can expect to accumulate over the future, starting from that state. The trial-and-error process leads to the well-known exploration versus exploitation trade-off. Indeed, to maximize the value function, the controller has to select actions (or policies) which have been tried before and which lead to high immediate reward, and most important, which lead to high long-term value. However, to discover these actions with high reward, the controller has to try as many different actions as needed. This trial versus application of control actions is the exploitation (application) versus exploration (trial) dilemma, which characterizes most of the model-free learning controllers. It is also worth noting that some reinforcement learning algorithms use a trial-and-error step not to learn the states-to-best actions mapping, but rather the model of the system. The model is then used for planning of the future control actions. In this book, we still refer to these types of control algorithms as model-free because the model is learned online, from "scratch," using direct interaction with the system. There are a lot of reinforcement learning methods available in the literature; they all use the same main ingredients mentioned previously, however, they differ in their algorithms, for example, in which way they estimate the long-term value function, and so on. This book is not about model-free adaptive control, nor is it about reinforcement learning methods, so we will leave it at that, and instead direct the reader to the references cited previously for a more detailed presentation of this topic.

These were only two concrete examples of model-free control algorithms; many more approaches have been proposed, for instance, evolutionary methods such as genetic algorithms, simulated annealing methods, and so on. One could also cite here the pure[3] neural network (NN), deep NN algorithms, see Prabhu and Garg (1996), Martinetz and Schulten (1993), Levine (2013), and Wang et al. (2016); and iterative learning control (ILC), see Bristow et al. (2006), Moore (1999), and Ahn et al. (2007).

We can now move on to the next section of this chapter, where we talk about adaptive methods more in line with the main topic of this book, namely, learning-based adaptive controllers.

---

[3]Pure in the sense that they do not use any physics-based model of the system.

2.5 LEARNING-BASED ADAPTIVE CONTROL

As we mentioned in Section 2.1, by learning-based controllers we mean controllers which are partly based on a physics–based model, and partly based on a model-free learning algorithm. Model-free learning is used to complement the physics–based model, and compensate for the uncertain or the missing part of the model. This compensation is done either directly by learning the uncertain part, or indirectly by tuning the controller to deal with the uncertainty.

Recently, there have been a lot of efforts in this "new" direction of adaptive control. One of the reasons behind this increasing interest is that the field of model-free learning has reached a certain maturity, which has led to good analysis and understanding of the main properties of the available model-free learning algorithms. The idea which we want to promote in this book, which is based on combining a model-based classical controller with a model-free learning algorithm, is attractive indeed. The reason is that by this combination, one could take advantage of the model-based design, with its stability characteristics, and add to it the advantages of the model-free learning, with its fast convergence and robustness to uncertainties. This combination is usually referred to as dual or modular design for adaptive control. In this line of research in adaptive control we can cite the following references: Wang and Hill (2010), Spooner et al. (2002), Zhang and Ordóñez (2012), Lewis et al. (2012), Wang et al. (2006), Guay and Zhang (2003), Vrabie et al. (2013), Koszaka et al. (2006), Frihauf et al. (2013), Haghi and Ariyur (2011, 2013), Jiang and Jiang (2013), Benosman and Atinc (2013a,c), Atinc and Benosman (2013), Modares et al. (2013), Benosman et al. (2014), Benosman (2014a,b,c), Benosman and Xia (2015), Gruenwald and Yucelen (2015), and Subbaraman and Benosman (2015).

For example, in the NN-based modular adaptive control design, the idea is to write the model of the system as a combination of a known part and an unknown part (a.k.a. the disturbance part). The NN is then used to approximate the unknown part of the model. Finally, a controller based on both the known and the NN estimate of the unknown part is determined to realize some desired regulation or tracking performance, see Wang and Hill (2010), Spooner et al. (2002), and Wang et al. (2006).

Let us report a simple second-order example to illustrate this idea. Consider the state-space model (in Brunowsky form)

$$\begin{cases} \dot{x}_1 = x_2, \\ \dot{x}_2 = f(x) + u, \end{cases} \qquad (2.66)$$

where f is an unknown smooth nonlinear scalar function of the state variables $x = (x_1,\ x_2)^T$, and u is a scalar control variable. The unknown part of the model f is approximated by an NN-based estimation $\hat{f}$ as follows:

$$\hat{f} = \hat{W}^T S(x), \tag{2.67}$$

where $\hat{W} = (\hat{w}_1, \ldots, \hat{w}_N)^T \in \mathbb{R}^N$ is the estimation of the weight vector of dimension N, which is the number of the NN nodes. The vector $S(x) = (s_1(x), \ldots, s_N(x))^T \in \mathbb{R}^N$ is the regressor vector, with s_i, $i = 1, \ldots, N$ representing some radial basis functions. Next, we consider the reference model

$$\begin{cases} \dot{x}_{ref_1} = x_{ref_2}, \\ \dot{x}_{ref_2} = f_{ref}(x), \end{cases} \tag{2.68}$$

where f_{ref} is a known nonlinear smooth function of the desired state trajectories $x_{ref} = (x_{ref_1},\ x_{ref_2})^T$. We assume that the f_{ref} is chosen such that the reference trajectories are uniformly bounded in time, and orbital, that is, repetitive motion, starting from any desired initial conditions $x_{ref}(0)$. Then a simple learning-based controller is

$$
\begin{aligned}
u &= -e_1 - c_1 e_2 - \hat{W}^T S(e) + \dot{v}, \\
e_1 &= x_1 - x_{ref_1}, \\
e_2 &= x_2 - v, \\
v &= -c_2 e_1 + x_{ref_2}, \\
\dot{v} &= -c_2(-c_2 e_1 + e_2) + f_{ref}(x_{ref}), \quad c_1,\ c_2 > 0, \\
\dot{\hat{W}} &= \Gamma(S(e)e_2 - \sigma \hat{W}), \quad \sigma > 0,\ \Gamma = \Gamma^T > 0.
\end{aligned}
\tag{2.69}
$$

This is a modular controller because the expression of the control u in Eq. (2.69) is dictated by the Brunowsky control form of the model (this is the known model-based information used in the control), and the remainder of the control is based on a model-free NN network estimating the unknown part of the model f. This controller has been proven to ensure uniform boundedness of the closed-loop signals, practical exponential convergence of the states' trajectories to a neighborhood of the reference orbital trajectories, and convergence of the regressor vector to its optimal value, that is, optimal estimation of the model uncertainty, see Wang and Hill (2010, Theorem 4.1). This is only a very simple example of NN-based learning-based adaptive controllers. More algorithms and analysis of these types of controllers can be found, for example, in Wang and Hill (2010) and Spooner et al. (2002), and references therein.

We also want to mention here some of the learning-based control methods based on ES algorithms. For instance, some specific ES-based control algorithms called numerical optimization-based ES algorithms, see Zhang and Ordóñez (2012), are used to optimize a desired performance cost function under the constraints of the system dynamics. The reason we cite this type of approach in this section, dedicated to learning-based adaptive controllers, is that these ES-based controllers do not assume the explicit knowledge of the cost function. Instead, these algorithms rely on the measurement of the cost function (and maybe its gradient) to generate a sequence of desired states which can lead the system to an optimal value for the performance cost function. To do so, the system's model (assumed known) is then used to design a model-based controller which forces the system's states to track (as close as possible) these desired states. So if we examine these algorithms closely, we see that they use a model-free step to optimize an unknown performance function, and then use a model-based step to guide the system's dynamics toward the optimal performance. Due to these two steps we classify these algorithms as learning-based adaptive (w.r.t. the unknown cost) controllers.

To clarify things, let us present a general formulation of a numerical optimization-based ES control. Let us consider the following nonlinear dynamics

$$\dot{x} = f(x, u), \tag{2.70}$$

where $x \in \mathbb{R}^n$ is the state vector, u is the control (assumed to be a scalar, to simplify the presentation), and f is a smooth known (possibly nonlinear) vector function. We associate with the system (2.70), a desired performance cost function $Q(x)$, where Q is a scalar smooth function of x. However, the explicit expression of Q as function of x (or u) is not known. In other words, the only available information is direct measurements of Q (and maybe its gradient). The goal is then to iteratively learn a control input which seeks the minimum of Q.

One of such extremum-seekers algorithms is the following:

Step 1: Initiate: $t_0 = 0$, $x(t_0) = x_0^s$ (chosen in $\mathbb{R}^n$), and the iteration step $k = 0$.

Step 2: Use a numerical optimization algorithm to generate x_{k+1}^s, based on the measurements $Q(x(t_k))$ (and if available $\nabla Q(x(t_k))$), s.t., $x_{k+1}^s = \text{argmin}(J(x(t_k)))$.

Step 3: Using the known model (2.70), design a state regulator u to regulate the actual state $x(t_{k+1})$ to the desired (optimal) state x^s_{k+1}, in a finite time δ_k, that is, $x(t_{k+1}) = x^s_{k+1}$, $t_{k+1} = t_k + \delta_k$.

Step 4: Increment the iterations index k to $k + 1$, and go back to Step 2.

Under some assumptions ensuring the existence of a global minimum, which is a stabilizable equilibrium point of Eq. (2.70), it has been shown that this type of algorithm converges to the minimum of the performance cost function, see Zhang and Ordóñez (2012, Chapters 3–5).

For example, if we consider the simple linear time-invariant model,

$$\dot{x} = Ax + Bu, \quad x \in \mathbb{R}^n, \ u \in \mathbb{R}. \tag{2.71}$$

We assume that the pair (A, B) is controllable. In this case, the previous numerical optimization-based ES algorithm reduces to the following:

Step 1: Initiate: $t_0 = 0$, $x(t_0) = x^s_0$ (chosen in $\mathbb{R}^n$), and the iteration step $k = 0$.

Step 2: Use a descent optimization algorithm to generate x^s_{k+1}, based on the measurements $Q(x(t_k))$ and $\nabla Q(x(t_k))$, s.t., $x^s_{k+1} = x(t_k) - \alpha_k \nabla Q(x(t_k))$, $\alpha_k > 0$.

Step 3: Choose a finite regulation time δ_k, and compute the model-based regulation control

$$u(t) = -B^T e^{A^T(t_{k+1}-t)} G^{-1}(\delta_k)(e^{A\delta_k}x(t_k) - x^s_{k+1}),$$

$$G(\delta_k) = \int_0^{\delta_k} e^{A\tau} BB^T e^{A^T\tau} d\tau,$$

for $t_k \le t \le t_{k+1} = t_k + \delta_k$.

Step 4: Test $\nabla Q(x(t_{k+1})) < \epsilon$, ($\epsilon > 0$, a convergence threshold), if yes End; otherwise, increment the iterations index k to $k + 1$, and go back to Step 2.

It has been shown in Zhang and Ordóñez (2012, Theorem 4.1.5) that this numerical optimization-based ES controller, under the assumption of existence of a stabilizable minimum, converges (globally and asymptotically) to the first-order stationary point of the cost function Q.

Many more ES-based adaptive controllers, which fall into the class of learning-based adaptive controllers, have been proposed in the literature. We cannot possibly present here all of these results; instead we refer the readers to, see Guay and Zhang (2003), Haghi and Ariyur (2011), Haghi and Ariyur (2013), Frihauf et al. (2013), Atinc and Benosman (2013),

Benosman (2014c), Benosman (2014b), Benosman (2014a), Benosman et al. (2014), Benosman and Xia (2015), and Subbaraman and Benosman (2015) and references therein, for some more examples on this topic. We also underline here that the work introduced in Atinc and Benosman (2013), Benosman (2014c), Benosman (2014b), Benosman (2014a), Benosman et al. (2014), Benosman and Xia (2015), and Subbaraman and Benosman (2015), are the main inspiration for this monograph, and will be discussed in more detail in the remaining chapters of this book.

To end this section, we want to mention some recent results in the field of learning-based adaptive control, namely, the reinforcement-learning (RL)-based adaptive controllers, see Koszaka et al. (2006), Lewis et al. (2012), and Vrabie et al. (2013). For instance, some of these results, introduced in Vrabie et al. (2013), called optimal adaptive controllers, are based on using RL value iteration algorithms to design adaptive controllers that learn the solutions to optimal control problems in real-time by observing real-time data.

Let us present one of these algorithms on the simple case of linear time-invariant continuous systems described by the model

$$\dot{x} = Ax + Bu, \tag{2.72}$$

with $x \in \mathbb{R}^n$, $u \in \mathbb{R}^m$, and the pair (A, B) assumed to be stabilizable. We underline here that the main assumption which makes this case classifiable as learning-based adaptive control is that the drift matrix A is not known, and the control matrix B is known. Thus the controller will be partly based on the model (based on B), and partly model-free (uses RL to compensate for the unknown part A). This model is then associated with a linear quadratic regulator (LQR)-type cost function of the form

$$V(u) = \int_{t_0}^{\infty} (x^T(\tau)R_1 x(\tau) + u^T(\tau)R_2 u(\tau))d\tau, \quad R_1 \geq 0, \ R_2 > 0, \tag{2.73}$$

where R_1 is chosen such that $(R_1^{1/2}, A)$ is detectable. The LQR optimal control is the controller satisfying

$$u^*(t) = \text{argmin}_{u(t)} V(u), \quad t \in [t_0, \infty[. \tag{2.74}$$

It is well known that, see Kailath (1980), the solution (in the nominal case with a known model), is given by

$$u^*(t) = -Kx(t),$$
$$K = R_2^{-1}B^T P, \tag{2.75}$$

where P is the solution to the famous algebraic Riccati equation

$$A^T P + PA - PBR_2^{-1}B^T P + Q = 0, \tag{2.76}$$

which has a unique semidefinite solution, under the detectability condition mentioned previously. However, the previous (classical) solution relies on the full knowledge of the model (2.72). In our case, we assumed that A was unknown, which requires some learning. One way to learn P, and thus learn the optimal control u^*, is based on an iterative RL algorithm called integral reinforcement learning policy iteration algorithm (IRL-PIA), see Vrabie et al. (2013, Chapter 3). The IRL-PIA is based on the iterative solution of the following equations:

$$x^T P_i x = \int_t^{t+T} x^T(\tau)(R_1 + K_i^T R_2 K_i)x(\tau)d\tau + x^T(t+T)P_i x(t+T),$$

$$K_{i+1} = R_2^{-1}B^T P_i, \quad i = 1, 2, \dots \tag{2.77}$$

where the initial gain K_1 is chosen such that $A - BK_1$ is stable. It has been proven in Vrabie et al. (2013, Theorem 3.4) that the policy iteration (2.77), if initiated from a stabilizing gain K_1, under the assumptions of stabilizability of (A, B) and detectability of $(R_1^{1/2}, A)$, converges to the optimal LQR solution u^* given by Eqs. (2.75), (2.76).

This was just a simple example to show the reader a case of RL-based adaptive controllers which rely partly on the model and partly on model-free learning iterations. The more challenging case of nonlinear systems has been also studied. For more details about this type of algorithm please refer to Vrabie et al. (2013), Lewis et al. (2012), and the references therein.

2.6 CONCLUSION

In this chapter we wanted to give an overview of the adaptive control field. This task turned out to be extremely difficult to fulfill, due to all the outstanding results which have been proposed during the past 70 years or so. To make the task more feasible we decided to decompose the field of adaptive control into three main streams: model-based, model-free, and learning-based adaptive control. Indeed, we defined model-based (classical) adaptive control as fully relying on some physics-based models of the system. On the opposite side, we defined model-free adaptive control as adaptive algorithms relying entirely on measurements data. Finally, as a hybrid approach, we defined learning-based adaptive control as adaptive algorithms using part of the system's model to design a model-based control, which is then complemented with a model-free learning algorithm to compensate for the unknown parts of the model. In each case, we *tried* to present some simple examples and cite some relevant references in each subfield. We do want to emphasize here the world *tried* because citing and presenting all the available results in one chapter is an unreachable goal. We do apologize for all the missing references, and hope that the chapter still gives a good overview of the main adaptive control results to the reader who can find more details in the specialized monographs cited in each subfield.

REFERENCES

Ahn, H.S., Chen, Y., Moore, K.L., 2007. Iterative learning control: brief survey and categorization. IEEE Trans. Syst. Man Cybern. 37 (6), 1099.

Ajami, A.F., 2005. Adaptive flight control in the presence of input constraints. Master's thesis, Virginia Polytechnic Institute and State University.

Ariyur, K.B., Krstic, M., 2002. Multivariable extremum seeking feedback: analysis and design. In: Proceedings of the Mathematical Theory of Networks and Systems, South Bend, IN.

Ariyur, K.B., Krstic, M., 2003. Real Time Optimization by Extremum Seeking Control. John Wiley & Sons, Inc., New York, NY, USA.

Astolfi, A., Karagiannis, D., Ortega, R., 2008. Nonlinear and Adaptive Control with Applications. Springer, London.

Astrom, K.J., 1983. Theory and applications of adaptive control—a survey. Automatica 19 (5), 471–486.

Astrom, K., Eykhoff, P., 1971. System identification—a survey. Automatica 7, 123–162.

Astrom, K.J., Wittenmark, B., 1995. A survey of adaptive control applications. In: IEEE, Conference on Decision and Control, pp. 649–654.

Astrom, K.J., Hagglund, T., Hang, C.C., Ho, W.K., 1994. Automatic tuning and adaptation for PID controllers—a survey. Control Eng. Pract. 1 (4), 699–714.

Atinc, G., Benosman, M., 2013. Nonlinear learning-based adaptive control for electromagnetic actuators with proof of stability. In: IEEE, Conference on Decision and Control, Florence, pp. 1277–1282.

Barkana, I., 2014. Simple adaptive control—a stable direct model reference adaptive control methodology—brief survey. Int. J. Adapt. Control Signal Process. 28, 567–603.

Bellman, R., 1957. Dynamic Programming. Princeton University Press, Princeton.

Bellman, R., 1961. Adaptive Process—A Guided Tour. Princeton University Press, Princeton.

Benosman, M., 2014a. Extremum-seeking based adaptive control for nonlinear systems. In: IFAC World Congress, Cape Town, South Africa, pp. 401–406.

Benosman, M., 2014b. Learning-based adaptive control for nonlinear systems. In: IEEE European Control Conference, Strasbourg, FR, pp. 920–925.

Benosman, M., 2014c. Multi-parametric extremum seeking-based auto-tuning for robust input-output linearization control. In: IEEE, Conference on Decision and Control, Los Angeles, CA, pp. 2685–2690.

Benosman, M., Atinc, G., 2013a. Multi-parametric extremum seeking-based learning control for electromagnetic actuators. In: IEEE, American Control Conference, Washington, DC, pp. 1914–1919.

Benosman, M., Atinc, G., 2013b. Nonlinear adaptive control of electromagnetic actuators. In: SIAM Conference on Control and Applications, pp. 29–36.

Benosman, M., Atinc, G., 2013c. Nonlinear learning-based adaptive control for electromagnetic actuators. In: IEEE, European Control Conference, Zurich, pp. 2904–2909.

Benosman, M., Atinc, G., 2015. Nonlinear adaptive control of electromagnetic actuators. IET Control Theory Appl., 258–269.

Benosman, M., Xia, M., 2015. Extremum seeking-based indirect adaptive control for nonlinear systems with time-varying uncertainties. In: IEEE, European Control Conference, Linz, Austria, pp. 2780–2785.

Benosman, M., Cairano, S.D., Weiss, A., 2014. Extremum seeking-based iterative learning linear MPC. In: IEEE Multi-conference on Systems and Control, pp. 1849–1854.

Bertsekas, D., Tsitsiklis, J., 1996. Neurodynamic Programming. Athena Scientific, Cambridge, MA.

Bristow, D.A., Tharayil, M., Alleyne, A.G., 2006. A survey of iterative learning control. IEEE Control Syst. 26 (3), 96–114.

Busonio, L., Babuska, R., Schutter, B.D., 2008. A comprehensive survey of multiagent reinforcement learning. IEEE Trans. Syst. Man Cybern. C: Appl. Rev. 38 (2), 156–172.

Busoniu, L., Babuska, R., De Schutter, B., Ernst, D., 2010. Reinforcement learning and dynamic programming using function approximators, Automation and Control Engineering. CRC Press, Boca Raton, FL.

Byrnes, C., Isidori, A., Willems, J.C., 1991. Passivity, feedback equivalence, and the global stabilization of minimum phase nonlinear systems. IEEE, Trans. Autom. Control 36 (11), 1228–1240.

Cao, C., Hovakimyan, N., 2008. Design and analysis of a novel L1 adaptive control architecture with guaranteed transient performance. IEEE Trans. Autom. Control 53 (2), 586–591.

Carr, N., 2014. The Glass Cage: Automation and us. W.W. Norton & Company, New York.

Drakunov, S., Ozguner, U., 1992. Optimization of nonlinear system output via sliding-mode approach. In: IEEE Workshop on Variable Structure and Lyapunov Control of Uncertain Dynamical Systems, University of Sheffield, UK.

Duarte, M.A., Narendra, K.S., 1989. Combined direct and indirect approach to adaptive control. IEEE Trans. Autom. Control 34 (10), 1071–1075.

Egardt, B., 1979. Stability of Adaptive Controllers. Springer-Verlag, Berlin.

Farahmand, A.M., 2011. Regularization in reinforcement learning. Ph.D. Thesis, University of Alberta.

Fekri, S., Athans, M., Pascoal, A., 2006. Issues, progress and new results in robust adaptive control. Int. J. Adapt. Control Signal Process. 20, 519–579.

Filatov, N.M., Unbehauen, H., 2000. Survey of adaptive dual control methods. IET Control Theory Appl. 147 (1), 118–128.

Flores-Perez, A., Grave, I., Tang, Y., 2013. Contraction based adaptive control for a class of nonlinearly parameterized systems. In: IEEE, American Control Conference, Washington, DC, pp. 2655–2660.

Fradkov, A.L., 1980. Speed-gradient scheme and its application in adaptive control problems. Autom. Remote Control 40 (9), 1333–1342.

Fradkov, A.L., 1994. Nonlinear adaptive control: regulation-tracking-oscillations. In: First IFAC Workshop: New Trends in Design of Control Systems, Smolenice, Slovakia, pp. 426–431.

Fradkov, A.L., Stotsky, A.A., 1992. Speed gradient adaptive control algorithms for mechanical systems. Int. J. Adapt. Control Signal Process. 6, 211–220.

Fradkov, A., Miroshnik, I., Nikiforov, V., 1999. Nonlinear and Adaptive Control of Complex Systems. Kluwer Academic Publishers, The Netherlands.

Fradkov, A., Ortega, R., Bastin, G., 2001. Semi-adaptive control of convexly parametrized systems with application to temperature regulation of chemical reactors. Int. J. Adapt. Control Signal Process. 15, 415–426.

Frihauf, P., Krstic, M., Basar, T., 2013. Finite-horizon LQ control for unknown discrete-time linear systems via extremum seeking. Eur. J. Control 19 (5), 399–407.

Giovanini, L., Benosman, M., Ordys, A., 2005. Adaptive control using multiple models switching and tuning. In: International Conference on Industrial Electronics and Control Applications, pp. 1–8.

Giovanini, L., Sanchez, G., Benosman, M., 2014. Observer-based adaptive control using multiple-models switching and tuning. IET Control Theory Appl. 8 (4), 235–247.

Goodwin, G.C., Sin, K.S., 1984. Adaptive Filtering Prediction and Control. Prentice-Hall, Englewood Cliffs, NJ.

Gregory, P.C., 1959. Proceedings of the self-adaptive flight control systems symposium. WADC Technical Report. Wright Air Development Centre, Ohio.

Gruenwald, B., Yucelen, T., 2015. On transient performance improvement of adaptive control architectures. Int. J. Control 88 (11), 2305–2315.

Guay, M., Zhang, T., 2003. Adaptive extremum seeking control of nonlinear dynamic systems with parametric uncertainties. Automatica 39, 1283–1293.

Guay, M., Dhaliwal, S., Dochain, D., 2013. A time-varying extremum-seeking control approach. In: IEEE, American Control Conference, pp. 2643–2648.

Haghi, P., Ariyur, K., 2011. On the extremum seeking of model reference adaptive control in higher-dimensional systems. In: IEEE, American Control Conference.

Haghi, P., Ariyur, K., 2013. Adaptive feedback linearization of nonlinear MIMO systems using ES-MRAC. In: IEEE, American Control Conference, pp. 1828–1833.

Hung, N.V.Q., Tuan, H.D., Narikiyo, T., Apkarian, P., 2008. Adaptive control for nonlinearly parameterized uncertainties in robot manipulators. IEEE Trans. Control Syst. Technol. 16 (3), 458–468.

Ilchmann, A., Ryan, E.P., 2008. High-gain control without identification: a survey. GAMM-Mitt. 31 (1), 115–125.

Ioannou, P., Sun, J., 2012. Robust Adaptive Control. Dover Publications, Mineola, NY.

Ioannou, P.A., Annaswamy, A.M., Narendra, K.S., Jafari, S., Rudd, L., Ortega, R., Boskovic, J., 2014. L1-adaptive control: stability, robustness, and interpretations. IEEE Trans. Autom. Control 59 (11), 3075–3080.

Isermann, R., 1982. Parameter adaptive control algorithms—a tutorial. Automatica 18 (5), 513–528.

Isidori, A., 1989. Nonlinear Control Systems, second ed., Communications and Control Engineering Series. Springer-Verlag, Berlin.

Jarvis, R.A., 1975. Optimization strategies in adaptive control: a selective survey. IEEE Trans. Syst. Man Cybern. 5 (1), 83–94.

Jiang, Z.P., Jiang, Y., 2013. Robust adaptive dynamic programming for linear and nonlinear systems: an overview. Eur. J. Control 19 (5), 417–425.

Jones, D.R., Perttunen, C.D., Stuckman, B.E., 1993. Lipschitzian optimization without the Lipschitz constant. J. Optim. Theory Appl. 79 (1), 157–181.

Kailath, T., 1980. Linear Systems. Prentice-Hall, Englewood Cliffs, NJ.

Karsenti, L., Lamnabhi-Lagarrigue, F., Bastin, G., 1996. Adaptive control of nonlinear systems with nonlinear parameterization. Syst. Control Lett. 27, 87–97.

Khong, S.Z., Nešić, D., Tan, Y., Manzie, C., 2013b. Unified frameworks for sampled-data extremum seeking control: global optimisation and multi-unit systems. Automatica 49 (9), 2720–2733.

Kormushev, P., Calinon, S., Caldwell, D.G., 2010. Robot motor skill coordination with EM-based reinforcement learning. In: IEEE/RSJ International Conference on Intelligent Robots and Systems, Taipei, Taiwan, pp. 3232–3237.

Koszaka, L., Rudek, R., Pozniak-Koszalka, I., 2006. An idea of using reinforcement learning in adaptive control systems. In: International Conference on Networking, International Conference on Systems and International Conference on Mobile Communications and Learning Technologies, 2006. ICN/ICONS/MCL 2006, p. 190.

Krasovskii, A.A., 1976. Optimal algorithms in problems of identification with an adaptive model. Avtom. Telemekh. 12, 75–82.

Krasovskii, A.A., Shendrik, V.S., 1977. A universal algorithm for optimal control of continuous processes. Avtomat. i Telemekh. 2, 5–13 (in Russian).

Krstic, M., 2000. Performance improvement and limitations in extremum seeking. Syst. Control Lett. 39, 313–326.

Krstić, M., Wang, H.H., 2000. Stability of extremum seeking feedback for general nonlinear dynamic systems. Automatica 36 (4), 595–601.

Krstic, M., Kanellakopoulos, I., Kokotovic, P., 1995. Nonlinear and Adaptive Control Design. John Wiley & Sons, New York.

Kumar, P.R., 1985. A survey of some results in stochastic adaptive control. SIAM J. Control Optim. 23 (3), 329–380.

Landau, I.D., 1979. Adaptive Control. Marcel Dekker, New York.

Landau, I.D., Lozano, R., M'Saad, M., Karimi, A., 2011. Adaptive control: Algorithms, analysis and applications, Communications and Control Engineering. Springer-Verlag, Berlin.

Lavretsky, E., 2009. Combined/composite model reference adaptive control. IEEE Trans. Autom. Control 54 (11), 2692–2697.

Leblanc, M., 1922. Sur lélectrification des chemins de fer au moyen de courants alternatifs de fréquence élevée. Revue Générale de lElectricité.

Levine, S., 2013. Exploring deep and recurrent architectures for optimal control. In: Neural Information Processing Systems (NIPS) Workshop on Deep Learning.

Lewis, F.L., Vrabie, D., Vamvoudakis, K.G., 2012. Reinforcement learning and feedback control: using natural decision methods to design optimal adaptive controllers. IEEE Control. Syst. Mag. 76–105, doi:10.1109/MCS.2012.2214134.

Liao, F., Lum, K.Y., Wang, J.L., Benosman, M., 2009. Adaptive nonlinear control allocation of non-minimum phase uncertain systems. In: IEEE, American Control Conference, St. Louis, MO, USA, pp. 2587–2592.

Liao, F., Lum, K.Y., Wang, J.L., Benosman, M., 2010. Adaptive control allocation for nonlinear systems with internal dynamics. IET Control Theory Appl. 4 (6), 909–922.

Liu, S.J., Krstic, M., 2014. Newton-based stochastic extremum seeking. Automatica 50 (3), 952–961.

Liu, S.J., Krstic, M., 2015. Stochastic averaging in discrete time and its applications to extremum seeking. IEEE Trans. Autom. Control 61 (1), 90–102.

Liu, X., Ortega, R., Su, H., Chu, J., 2010. Immersion and invariance adaptive control of nonlinearly parameterized nonlinear systems. IEEE Trans. Autom. Control 55 (9), 2209–2214.

Ljung, L., Gunnarsson, S., 1990. Adaptation and tracking in system identification—a survey. Automatica 26 (1), 7–21.

Martinetz, T., Schulten, K., 1993. A neural network for robot control: cooperation between neural units as a requirement for learning. Comput. Electr. Eng. 19 (4), 315–332.

Modares, R., Lewis, F., Yucelen, T., Chowdhary, G., 2013. Adaptive optimal control of partially-unknown constrained-input systems using policy iteration with experience replay. In: AIAA Guidance, Navigation, and Control Conference, Boston, MA, doi: 10.2514/6.2013-4519.

Moore, K.L., 1999. Iterative learning control: an expository overview. In: Applied and Computational Control, Signals, and Circuits. Springer, New York, pp. 151–214.

Narendra, K.S., Annaswamy, A.M., 1989. Stable Adaptive Systems. Prentice-Hall, Englewood Cliffs, NJ.

Narendra, K.S., Balakrishnan, J., 1997. Adaptive control using multiple models. IEEE Trans. Autom. Control 42 (2), 171–187.

Narendra, K.S., Driollet, O.A., Feiler, M., George, K., 2003. Adaptive control using multiple models, switching and tuning. Int. J. Adapt. Control Signal Process. 17, 87–102.

Nesic, D., 2009. Extremum seeking control: convergence analysis. Eur. J. Control 15 (3-4), 331–347.

Noase, W., Tan, Y., Nesic, D., Manzie, C., 2011. Non-local stability of a multi-variable extremum-seeking scheme. In: IEEE, Australian Control Conference, pp. 38–43.

Ortega, R., Fradkov, A., 1993. Asymptotic stability of a class of adaptive systems. Int. J. Adapt. Control Signal Process. 7, 255–260.

Ortega, R., Tang, Y., 1989. Robustness of adaptive controllers—a survey. Automatica 25 (5), 651–677.

Park, B.S., Lee, J.Y., Park, J.B., Choi, Y.H., 2012. Adaptive control for input-constrained linear systems. Int. J. Control Autom. Syst. 10 (5), 890–896.

Prabhu, S.M., Garg, D.P., 1996. Artificial neural network based robot control: an overview. J. Intell. Robot. Syst. 15 (4), 333–365.

Rojas, O.J., Goodwin, G.C., Desbiens, A., 2002. Study of an adaptive anti-windup strategy for cross-directional control systems. In: IEEE, Conference on Decision and Control, pp. 1331–1336.

Rotea, M., 2000. Analysis of multivariable extremum seeking algorithms. In: Proceedings of the American Control Conference, vol. 1. IEEE, pp. 433–437.

Rusnak, I., Weiss, H., Barkana, I., 2014. Improving the performance of existing missile autopilot using simple adaptive control. Int. J. Robust Nonlinear Control 28, 732–749.

Sastry, S., Bodson, M., 2011. Adaptive Control: Stability, Convergence and Robustness. Dover Publications, Mineola.

Scheinker, A., 2013. Simultaneous stabilization and optimization of unknown, time-varying systems. In: American Control Conference (ACC), 2013, pp. 2637–2642.

Scheinker, A., Krstic, M., 2013. Maximum-seeking for CLFs: universal semiglobally stabilizing feedback under unknown control directions. IEEE Trans. Autom. Control 58, 1107–1122.

Seborg, D.E., Edgar, T.F., Shah, S.L., 1986. Adaptive control strategies for process control: a survey. AIChE J. 32 (6), 881–913.

Seron, M., Hill, D., Fradkov, A., 1995. Nonlinear adaptive control of feedback passive systems. Automatica 31 (7), 1053–1057.

Slotine, J., Li, W., 1991. Applied Nonlinear Control, Prentice-Hall International Edition. Prentice-Hall, Englewood Cliffs, NJ, pp. 68–73 .

Slotine, J.J.E., Li, W., 1989. Composite adaptive control of robot manipulators. Automatica 25 (4), 509–519.

Spooner, J.T., Maggiore, M., Ordonez, R., Passino, K.M., 2002. Stable adaptive control and estimation for nonlinear systems. Wiley-Interscience, New York.

Sragovich, V., 2006. Mathematical theory of adaptive control. In: Interdisciplinary Mathematical Sciences, vol. 4. World Scientific, Singapore, translated by: I.A. Sinitzin.

Subbaraman, A., Benosman, M., 2015. Extremum seeking-based iterative learning model predictive control (ESILC-MPC), Tech. rep., arXiv:1512.02627v1 [cs.SY].

Sutton, R.S., Barto, A.G., 1998. Reinforcement Learning: An Introduction. MIT Press, Cambridge, MA.

Szepesvári, C., 2010. Algorithms for Reinforcement Learning. Morgan & Claypool Publishers, California, USA.

Tan, Y., Li, Y., Mareels, I., 2013. Extremum seeking for constrained inputs. IEEE Trans. Autom. Control 58 (9), 2405–2410.

Tan, Y., Nesic, D., Mareels, I., 2006. On non-local stability properties of extremum seeking control. Automatica 42, 889–903.

Tan, Y., Nesic, D., Mareels, I., 2008. On the dither choice in extremum seeking control. Automatica 44, 1446–1450.

Tao, G., 2003. Adaptive control design and analysis. Hoboken, NJ: John Wiley and Sons.

Tao, G., 2014. Mutlivariable adaptive control: A survey. Automatica 50 (2014), 2737–2764.

Thorndike, E.L., 1911. Animal Intelligence; Experimental Studies, The Animal Behavior Series. The Macmillan Company, New York.

Tsypkin, Y.Z., 1971. Adaptation and Learning in Automatic Systems. Academic Press, New York.

Tsypkin, Y.Z., 1975. Foundations of the Theory of Learning Systems. Academic Press, New York.

Vrabie, D., Vamvoudakis, K., Lewis, F.L., 2013. Optimal Adaptive Control and Differential Games by Reinforcement Learning Principles, IET Digital Library.

Wang, C., Hill, D.J., 2010. Deterministic Learning Theory for Identification, Recognition, and Control. CRC Press, Boca Raton, FL.

Wang, C., Hill, D.J., Ge, S.S., Chen, G., 2006. An ISS-modular approach for adaptive neural control of pure-feedback systems. Automatica 42 (5), 723–731.

Wang, Z., Liu, Z., Zheng, C., 2016. Qualitative analysis and control of complex neural networks with delays, Studies in Systems, Decision and Control, vol. 34. Springer-Verlag, Berlin/Heidelberg.

Yakubovich, V., 1969. Theory of adaptive systems. Sov. Phys. Dokl. 13, 852–855.

Ye, M., Hu, G., 2013. Extremum seeking under input constraint for systems with a time-varying extremum. In: IEEE, Conference on Decision and Control, pp. 1708–1713.

Yeh, P.C., Kokotovit, P.V., 1995. Adaptive tracking designs for input-constrained linear systems using backstepping. In: IEEE, American Control Conference, pp. 1590–1594.

Zhang, C., Ordóñez, R., 2012. Extremum-Seeking Control and Applications: A Numerical Optimization-Based Approach. Springer, New York.

CHAPTER 3

Extremum Seeking-Based Iterative Feedback Gains Tuning Theory

3.1 INTRODUCTION

Nowadays, feedback controllers are used in a variety of systems. There are several types of feedback, for example, state feedback versus output feedback, linear versus nonlinear, and so on. However, one common characteristic of all available feedback controllers is the fact that they all rely on some "well-chosen" feedback gains. The selection of these feedback gains is often done based on some desired performance. For instance, the gains can be chosen in such a way to minimize overshoot of a linear closed-loop systems. Settling time can be another performance target; minimizing a given finite time or asymptotic state or output tracking error can be of interest in many applications as well.

Over the past years, there has been a myriad of results about feedback gains tuning. Maybe one of the most famous and widely taught techniques, is the Ziegler-Nichols rules for proportional-integral-derivative (PID) gains tuning for linear systems (Ziegler and Nichols, 1942). However, such rules apply for the particular class of linear systems, under linear PID feedback, and are considered heuristic in nature. For more general cases of models and controllers, and a more systematic or autonomous way of tuning feedback gains, the control community started looking at an iterative procedure to auto-tune feedback gains for closed-loop systems.

Indeed, in the seminal paper by Hjalmarsson et al. (1994), the authors introduced the idea that feedback controllers' parameters could be tuned iteratively to compensate for model uncertainties, and that the tuning could be based on measurements which are directly obtained from the system. This idea of iterative control tuning lead to the iterative feedback tuning (IFT) research field, where the goal is to iteratively auto-tune feedback gains of closed-loop systems, based on the online optimization of a well-defined performance cost function.

Learning-Based Adaptive Control
http://dx.doi.org/10.1016/B978-0-12-803136-0.00003-8
55

There have been a lot of results about IFT in the past 20 years, and it is not the purpose of this chapter to survey all the existing papers in the field. However, the existing results are mainly dedicated to linear systems controlled with linear feedbacks, for example, Lequin et al. (2003), Hjalmarsson (2002), Killingsworth and Kristic (2006), Koszalka et al. (2006), Hjalmarsson et al. (1998), and Wang et al. (2009). Based on these IFT algorithms for linear systems, some extensions to nonlinear systems have been studied. For instance in Hjalmarsson (1998), the author studied the case of discrete nonlinear systems controlled with linear time-invariant output feedback. The effect of IFT algorithms developed originally for linear systems, was studied on nonlinear systems by assuming local Taylor approximation of the nonlinear dynamics. However, the full analysis of the feedback loop, that is, IFT merged with the linear controller and the nonlinear dynamics was not reported in this paper. Other feedback gains iterative tuning algorithms were developed for nonlinear systems in Sjoberg and Agarwal (1996), DeBruyne et al. (1997), and Sjoberg et al. (2003). The algorithms developed in these papers first assume that the closed-loop input and output signals remain bounded during the gains tuning, and then rely on the numerical estimation of the gradient of a given cost function with respect to the controller gains, which necessitates running the system $n + 2$ times if the dimension of the tuned gain vector is n. This obviously can be a limiting factor if the number of tuned parameters is large.

In this chapter, we propose to study the problem of gains auto-tuning, in the general setting of uncertain nonlinear systems, with a rigorous stability analysis of the full system, that is, learning algorithm merged with the nonlinear controller, and the nonlinear uncertain system (refer to Benosman and Atinc (2013) and Benosman (2014), for preliminary results). We consider here a particular class of nonlinear systems, namely, nonlinear models affine in the control input, which are linearizable via static-state feedback. The types of uncertainties considered here are bounded additive model uncertainties, with a known upper-bound function. We propose a simple modular iterative gains tuning controller in the sense that we first design a robust controller, based on the classical input-output linearization method, merged with a Lyapunov reconstruction-based control, see Khalil (1996) and Benosman and Lum (2010). This robust controller ensures uniform boundedness of the tracking errors and their convergence to a given invariant set; that is, it ensures Lagrange stability of the system. Next, in a second phase we add a multiparametric extremum-seeking

(MES) algorithm to iteratively auto-tune the feedback gains of the robust controller. The extremum seeker optimizes a desired system performance, which is formulated in terms of a performance cost function.

One point worth mentioning at this stage is that compared to model-free pure MES-based controllers, the MES-based IFT control has a different goal. Indeed, the available pure MES-based controllers are meant for output or state regulation, that is, solving a static optimization problem. On the contrary, here we propose to use MES to complement a model-based nonlinear control to auto-tune its feedback gains. This means that the control goal, that is, state or output trajectory tracking, is handled by the model-based controller. The MES algorithm is used to improve the tracking performance of the model-based controller, and once the MES algorithm has converged, one can carry on using the nonlinear model-based feedback controller alone, without need of the MES algorithm. In other words, the MES algorithm is used here to replace the manual feedback gains tuning of the model-based controller, which is often done in real life by some type of trial-and-error tests.

It is also worth underlying here that the proposed MES-based nonlinear IFT method differs from the existing model-free iterative learning control (ILC) algorithm by two main points: First, the proposed method aims at auto-tuning a given vector of feedback gains associated with a nonlinear model-based robust controller. Thus, once the gains are tuned, the optimal gains obtained by the MES tuning algorithm can be used in the sequel without need of the MES algorithm. Second, the available model-free ILC algorithms do not require any knowledge about the controlled system. In other words, ILC in essence is a model-free control which does not need any knowledge of the system's physics. This can be appealing when the model of the system is hard to obtain; however, it comes at the expense of a large number of iterations needed to learn all the dynamics of the system (although indirectly via the learning of a given optimal feedforward control signal). In these conditions, we believe that our approach is faster in terms of the number of iterations needed to improve the overall performance of the closed-loop system. Indeed, our approach is based on the idea of using all the available information about the system's model to design in a first phase a model-based controller, and then in a second phase improve the performance of this controller by tuning its gains to compensate for the unknown or the uncertain part of the model. Because unlike the complete model-free ILC, we do not start from scratch, and do use some knowledge about the system's model, we expect to converge to an optimal performance faster than the model-free ILC algorithms.

This chapter is organized as follows: First, some notations and definitions are recalled in Section 3.2. Next, we present the class of systems studied here, and formulate the control problem in Section 3.3. The proposed control approach, together with its stability analysis, is presented in Section 3.4. Section 3.5 is dedicated to the application of the controller to two mechatronics examples, namely, an electromagnetic actuator and a two-link manipulator robot. Finally, the chapter ends with a summarizing conclusion, and a discussion of some open problems in Section 3.6.

3.2 BASIC NOTATIONS AND DEFINITIONS

Throughout this chapter we use $\|.\|$ to denote the Euclidean norm; that is, for $x \in \mathbb{R}^n$ we have $\|x\| = \sqrt{x^T x}$. We use the notations $\mathrm{diag}\{m_1, \ldots, m_n\}$ for $n \times n$ diagonal matrix of diagonal elements m_i's, $z(i)$ denotes the ith element of the vector z. We use $(\dot{.})$ for the short notation of time derivative, and $f^{(r)}(t)$ for $\frac{d^r f(t)}{dt^r}$. $\mathrm{Max}(V)$ denotes the maximum element of a vector V, and $\mathrm{sgn}(.)$ denotes for the sign function. The Frobenius norm of a matrix $A \in \mathbb{R}^{m \times n}$, with elements a_{ij}, is defined as $\|A\|_F \triangleq \sqrt{\sum_{i=1}^{n} \sum_{j=1}^{n} |a_{ij}|^2}$.

We denote by $\mathcal{C}^k$ functions that are k times differentiable, and by $\mathcal{C}^\infty$ a smooth function. A function is said to be analytic in a given set if it admits a convergent Taylor series approximation in some neighborhood of every point of the set. An impulsive dynamical system is said to be well posed if it has well-defined distinct resetting times, admits a unique solution over a finite forward time interval, and does not exhibit any Zeno solutions, that is, an infinitely many resetting of the system in finite time interval (Haddad et al., 2006). Finally, in the sequel, when we talk about error trajectories boundedness, we mean uniform boundedness as defined in Khalil (1996, p. 167, Definition 4.6) for nonlinear continuous systems, and in Haddad et al. (2006, p. 67, Definition 2.12) for time-dependent impulsive dynamical systems (cf. Chapter 1).

3.3 PROBLEM FORMULATION

3.3.1 Class of Systems

We consider here affine uncertain nonlinear systems of the form

$$\dot{x} = f(x) + \Delta f(x) + g(x)u, \quad x(0) = x_0,$$
$$y = h(x), \tag{3.1}$$

where $x \in \mathbb{R}^n$, $u \in \mathbb{R}^{n_a}$, and $y \in \mathbb{R}^m (n_a \geq m)$ represent the state, the input, and the controlled output vectors, respectively, x_0 is a given finite initial condition, $\Delta f(x)$ is a vector field representing additive model uncertainties. The vector fields f, Δf, columns of g, and function h satisfy the following assumptions.

Assumption 3.1. $f : \mathbb{R}^n \to \mathbb{R}^n$ *and the columns of* $g : \mathbb{R}^n \to \mathbb{R}^{n \times n_a}$ *are* $\mathcal{C}^\infty$ *vector fields on a bounded set* $X \subset \mathbb{R}^n$, $h(.)$ *is a* $\mathcal{C}^\infty$ *function on* X, *and the vector field* $\Delta f(.)$ *is* $\mathcal{C}^1$ *on* X.

Assumption 3.2. *System* (3.1) *has a well-defined (vector) relative degree* $\{r_1, \ldots, r_m\}$ *at each point* $x^0 \in X$, *and the system is linearizable, that is,* $\sum_{i=1}^{i=m} r_i = n$ *(cf. Chapter 1).*

Assumption 3.3. *The uncertainty vector function* $\Delta f(.)$ *is s.t.* $\|\Delta f(x)\| \leq d(x)$ $\forall x \in X$, *where* $d : X \to \mathbb{R}$ *is a smooth nonnegative function.*

Assumption 3.4. *The desired output trajectories* y_{id} *are smooth functions of time, relating desired initial points* y_{i0} *at* $t = 0$ *to desired final points* y_{if} *at* $t = t_f$, *and s.t.* $y_{id}(t) = y_{if}$, $\forall t \geq t_f$, $t_f > 0$, $i \in \{1, \ldots, m\}$.

3.3.2 Control Objectives

Our objective is to design a feedback controller $u(x, K)$, which ensures for the uncertain model (3.1) uniform boundedness of the output tracking error, and for which the stabilizing feedback gains vector K is iteratively auto-tuned online, to optimize a desired performance cost function.

We stress here that the goal of the gains' auto-tuning is not stabilization, but rather performance optimization. To achieve this control objective, we proceed as follows: we design a robust controller that ensures boundedness of the tracking error dynamics, and we combine it with a model-free learning algorithm to iteratively auto-tune the feedback gains of the controller, and optimize online a desired performance cost function.

In the next section we present a two-step design of a controller targeting the objectives stated previously.

3.4 EXTREMUM SEEKING-BASED ITERATIVE GAIN TUNING FOR INPUT-OUTPUT LINEARIZATION CONTROL

3.4.1 Step One: Robust Control Design

Under Assumption 3.2 and nominal conditions, that is, $\Delta f = 0$, system (3.1) can be written as, see Isidori (1989)

$$y^{(r)}(t) = b(\xi(t)) + A(\xi(t))u(t), \qquad (3.2)$$

where

$$y^{(r)}(t) \triangleq (y_1^{(r_1)}(t), \ldots, y_m^{(r_m)}(t))^T,$$
$$\xi(t) = (\xi^1(t), \ldots, \xi^m(t))^T, \tag{3.3}$$
$$\xi^i(t) = (y_i(t), \ldots, y_i^{(r_i-1)}(t)), \quad 1 \le i \le m,$$

and b, A write as functions of f, g, h, and A is nonsingular in X (Isidori, 1989, pp. 234–288).

At this point we introduce one more assumption.

Assumption 3.5. *We assume that the additive uncertainties Δf in Eq. (3.1) appear as additive uncertainties in the linearized models (3.2), (3.3) as follows:*

$$y^{(r)} = b(\xi) + \Delta b(\xi) + A(\xi)u, \tag{3.4}$$

where Δb is C^1 on $\tilde{X}$, and s.t. $\|\Delta b(\xi)\| \le d_2(\xi) \; \forall \xi \in \tilde{X}$, where $d_2 : \tilde{X} \to \mathbb{R}$ is a smooth nonnegative function, and $\tilde{X}$ is the image of the set X by the diffeomorphism $x \to \xi$ between the states of Eqs. (3.1), (3.2).

First, if we consider the nominal model (3.2), we can define a virtual input vector v as

$$b(\xi(t)) + A(\xi(t))u(t) = v(t). \tag{3.5}$$

Combining Eqs. (3.2), (3.5), we obtain the linear (virtual) input-output mapping

$$y^{(r)}(t) = v(t). \tag{3.6}$$

Based on the linear system (3.6), we write the stabilizing output feedback for the nominal system (3.4) with $\Delta b(\xi) = 0$, as

$$u_{\text{nom}} = A^{-1}(\xi)(v_s(t,\xi) - b(\xi)), \quad v_s = (v_{s1}, \ldots, v_{sm})^T$$
$$v_{si}(t,\xi) = y_{i_d}^{(ri)} - K_{ri}^i(y_i^{(ri-1)} - y_{i_d}^{(ri-1)}) - \cdots - K_1^i(y_i - y_{i_d}), i \in \{1, \ldots, m\}. \tag{3.7}$$

Denoting the tracking error vector as $e_i(t) = y_i(t) - y_{i_d}(t)$, we obtain the tracking error dynamics

$$e_i^{(r_i)}(t) + K_{r_i}^i e_i^{(r_i-1)}(t) + \cdots + K_1^i e_i(t) = 0, \; i = 1, \ldots, m, \tag{3.8}$$

and by tuning the gains K_j^i, $i = 1, \ldots, m$, $j = 1, \ldots, r_i$ such that all the polynomials in Eq. (3.8) are Hurwitz, we obtain global asymptotic convergence of the tracking errors $e_i(t)$, $i = 1, \ldots, m$, to zero. To formalize this condition let us state the following assumption.

Assumption 3.6. *We assume that there exists a nonempty set $\mathcal{K}$ of again K_j^i, $i = 1, \ldots, m$, $j = 1, \ldots, r_i$, such that the polynomials (3.8) are Hurwitz.*

Remark 3.1. *Assumption 3.6 is well known in the input-output lineariza-tion control literature. It simply states that we can find gains that stabilize the polynomials (3.8), which can be done, for example, by pole placements (see Section 3.5.1 for some examples).*

Next, if we consider that $\Delta b(\xi) \neq 0$ in Eq. (3.4), the global asymptotic stability of the error dynamics will not be guaranteed anymore due to the additive error vector $\Delta b(\xi)$, we then choose to use Lyapunov reconstruction technique (e.g., Benosman and Lum (2010)) to obtain a controller ensuring practical stability of the tracking error. This controller is presented in the following theorem.

Theorem 3.1. *Consider the system (3.1) for any $x_0 \in \mathbb{R}^n$, under Assumptions 3.1–3.6, with the feedback controller*

$$u = A^{-1}(\xi)(v_s(t, \xi) - b(\xi)) - A^{-1}(\xi)\frac{\partial V}{\partial \tilde{z}}^T k\, d_2(e),\ k > 0,$$

$$v_s = (v_{s1}, \ldots, v_{sm})^T,$$

$$v_{si}(t, \xi) = y_{i_d}^{(r_i)} - K_{r_i}^i(y_i^{(r_i-1)} - y_{i_d}^{(r_i-1)}) - \cdots - K_1^i(y_i - y_{i_d}) \qquad (3.9)$$

where $K_j^i \in \mathcal{K}$, $j = 1, \ldots, r_i$, $i = 1, \ldots, m$, and $V = z^T P z$, $P > 0$ such that $P\tilde{A} + \tilde{A}^T P = -I$, with $\tilde{A}$ being an $n \times n$ matrix defined as

$$\tilde{A} = \begin{pmatrix}
0, 1, 0, \ldots\ldots\ldots\ldots\ldots\ldots\ldots\ldots, 0 \\
0, 0, 1, 0, \ldots\ldots\ldots\ldots\ldots\ldots\ldots, 0 \\
\ddots \\
-K_1^1, \ldots, -K_{r1}^1, 0, \ldots\ldots\ldots\ldots\ldots, 0 \\
\ddots \\
0, \ldots\ldots\ldots\ldots\ldots, 0, 1, 0, \ldots\ldots, 0 \\
0, \ldots\ldots\ldots\ldots\ldots, 0, 0, 1, \ldots\ldots, 0 \\
\ddots \\
0, \ldots\ldots\ldots, 0, -K_1^m, \ldots\ldots, -K_{rm}^m
\end{pmatrix}, \qquad (3.10)$$

and $z = (z^1, \ldots, z^m)^T$, $z^i = (e_i, \ldots, e_i^{r_i-1})$, $i = 1, \ldots, m$, $\tilde{z} = (z^1(r_1), \ldots, z^m(r_m))^T \in \mathbb{R}^m$. Then, the vector z is uniformly bounded and reached the positive invariant set $S = \{z \in \mathbb{R}^n|\ 1 - k\ \|\frac{\partial V}{\partial \tilde{z}}\| \geq 0\}$.

Proof. We know from the previous discussion that for the system (3.1), under Assumptions 3.1, 3.2, and 3.6, in the nominal case, $\Delta f = 0$, the control (3.7) globally asymptotically (exponentially) stabilizes the linear error dynamic (3.8), which by the classical results in Khalil (1996, pp. 135–136), leads to the existence of a Lyapunov function $V = z^T P z$ s.t. the time derivative of V along the nominal system (3.1) (with $\Delta f = 0$), under the control law u_{nom} given by Eq. (3.7), satisfies

$$\dot{V}|_{(3.1),\Delta f=0} \leq -\|z\|^2,$$

where $z = (z^1, \ldots, z^m)^T$, $z^i = (e_i, \ldots, e_i^{r_i-1})$, $i = 1, \ldots, m$, and $P > 0$ is the unique solution of the Lyapunov equation $P\tilde{A} + \tilde{A}^T P = -I$, wherein $\tilde{A}$ given by Eq. (3.10) has been obtained by rewriting the error dynamic in the control canonical form. Now, we will use the technique of Lyapunov reconstruction from nonlinear robust control, see Benosman and Lum (2010), to obtain the full controller (3.9). Indeed, if we compute the time derivative of V along the uncertain model (3.1), under Assumptions 3.3 and 3.5, and considering the augmented control law $u = u_{\text{nom}} + u_{\text{robust}}$, we obtain

$$\dot{V}|_{(3.1),\Delta f \neq 0} \leq -\|z\|^2 + \frac{\partial V}{\partial \tilde{z}}.(A\, u_{\text{robust}} + \Delta b), \qquad (3.11)$$

where $\tilde{z} = (z^1(r_1), \ldots, z^m(r_m))^T \in \mathbb{R}^m$.
Next, if we define u_{robust} as

$$u_{\text{robust}} = -A^{-1}(\xi)\frac{\partial V}{\partial \tilde{z}}^T k\, d_2(e), \quad k > 0 \qquad (3.12)$$

substituting Eq. (3.12) in Eq. (3.11), we obtain

$$\dot{V}|_{(3.1),\Delta f \neq 0} \leq -\|z\|^2 - \left\|\frac{\partial V}{\partial \tilde{z}}\right\|^2 k\, d_2(e) + \frac{\partial V}{\partial \tilde{z}}\Delta b$$

$$\leq -\|z\|^2 - \left\|\frac{\partial V}{\partial \tilde{z}}\right\|^2 k\, d_2(e) + \left\|\frac{\partial V}{\partial \tilde{z}}\right\| d_2 \qquad (3.13)$$

$$\leq \left(1 - k\left\|\frac{\partial V}{\partial \tilde{z}}\right\|\right)\left\|\frac{\partial V}{\partial \tilde{z}}\right\| d_2$$

which proves that V is decreasing as long as $1 - k\left\|\frac{\partial V}{\partial \tilde{z}}\right\| < 0$, until the error vector enters the positive invariant set $S = \{z \in \mathbb{R}^n | 1 - k\left\|\frac{\partial V}{\partial \tilde{z}}\right\| \geq 0\}$, which implies boundedness of V, and equivalently uniform boundedness of $\|z\|$ (which can be directly obtained via the inequality $\lambda_{\min}(P)\|z\|^2 \leq V(z)$, e.g., Hale (1977)).

Remark 3.2. *In the proof of Theorem 3.1, we use a smooth control term* u_{robust} *given by Eq. (3.12); however, we could use a nonsmooth control by choosing* $u_{\text{robust}} = -A^{-1}\text{sgn}(\frac{\partial V}{\partial z}_{\text{ind}})' k \; d_2(e)$. *Indeed, this controller is well known to compensate for bounded uncertainties, and would lead to an asymptotic stability result for the tracking error dynamics, but it is discontinuous and thus not advisable for real applications. Its regularization is often done by replacing the sign function by a saturation function, see for example, Benosman and Lum (2010), which leads to practical stability results similar to the one obtained with the proposed* u_{robust} *term in Theorem 3.1.*

3.4.2 Step Two: Iterative Auto-Tuning of the Feedback Gains

In Theorem 3.1, we showed that the robust controller (3.9) leads to bounded tracking errors attracted to the invariant set S for a given choice of the feedback gains K_j^i, $j = 1, \ldots, ri$, $i = 1, \ldots, m$. Next, to iteratively auto-tune the feedback gains of Eq. (3.9), we define a desired cost function, and use an MES algorithm to iteratively auto-tune the gains and minimize the performance cost function. We first denote the cost function to be minimized as $Q(z(\beta))$, where β represents the optimization variables vector, defined as

$$\beta = (\delta K_1^1, \ldots, \delta K_{r1}^1, \ldots, \delta K_1^m, \ldots, \delta K_{rm}^m, \delta k)^T \tag{3.14}$$

such that the updated feedback gains write as

$$K_j^i = K_{j-\text{nominal}}^i + \delta K_j^i, \; j = 1, \ldots, ri, \; i = 1, \ldots, m,$$
$$k = k_{\text{nominal}} + \delta k, \quad k_{\text{nominal}} > 0, \tag{3.15}$$

where $K_{j-\text{nominal}}^i$, $j = 1, \ldots, ri$, $i = 1, \ldots, m$, are the nominal initial values of the feedback gains chosen such that Assumption 3.6 is satisfied.

Remark 3.3. *The choice of the cost function Q is not unique. For instance, if the system's tracking performance at specific time instants* It_f, $I \in \{1, 2, 3, \ldots\}$ *is important for the targeted application (see the electromagnetic actuator example presented in Section 3.5.1), one can choose Q as*

$$Q(z(\beta)) = z^T(It_f)C_1 z(It_f), \quad C_1 > 0. \tag{3.16}$$

If other performances need to be optimized over a finite time interval, for instance a combination of a tracking performance and a control power performance, then one can choose, for example, the cost function

$$Q(z(\beta)) = \int_{(I-1)t_f}^{It_f} z^T(t)C_1 z(t)\,dt + \int_{(I-1)t_f}^{It_f} u^T(t)C_2 u(t)\,dt, \tag{3.17}$$

where $I \in \{1, 2, 3, \ldots\}$, C_1, $C_2 > 0$. *The gains variation vector β is then used to minimize the cost function Q over the learning iterations $I \in \{1, 2, 3, \ldots\}$.*

Following the MES theory, see Ariyur and Krstic (2002), the variations of the gains are computed using the algorithm

$$\dot{x}_{K_j^i} = a_{K_j^i} \sin\left(\omega_{K_j^i} t - \frac{\pi}{2}\right) Q(z(\beta)),$$

$$\delta \hat{K}_j^i(t) = x_{K_j^i}(t) + a_{K_j^i} \sin\left(\omega_{K_j^i} t + \frac{\pi}{2}\right), \; j = 1, \ldots, ri, \; i = 1, \ldots, m,$$

$$\dot{x}_k = a_k \sin\left(\omega_k t - \frac{\pi}{2}\right) Q(z(\beta)), \tag{3.18}$$

$$\delta \hat{k}(t) = x_k(t) + a_k \sin\left(\omega_k t + \frac{\pi}{2}\right),$$

where $a_{K_j^i}$, $j = 1, \ldots, ri$, $i = 1, \ldots, m$, a_k are positive tuning parameters, and

$$\omega_1 + \omega_2 \neq \omega_3, \text{ for } \omega_1 \neq \omega_2 \neq \omega_3,$$

$$\forall \omega_1, \omega_2, \omega_3 \in \{\omega_{K_j^i}, \omega_k, j = 1, \ldots, ri, \; i = 1, \ldots, m\}, \tag{3.19}$$

with $\omega_i > \omega^*$, $\forall \omega_i \in \{\omega_{K_j^i}, \omega_k, \; j = 1, \ldots, ri, \; i = 1, \ldots, m\}$, ω^* large enough.

To study the stability of the learning-based controller, that is, controller (3.9), with the varying gains (3.15), (3.18), we first need to introduce some additional assumptions.

Assumption 3.7. *We assume that the cost function Q has (at least) a local minimum at β^*.*

Assumption 3.8. *We consider that the initial gain vector β is sufficiently close to the optimal gain vector β^*.*

Assumption 3.9. *The cost function is analytic and its variation with respect to the gains is bounded in the neighborhood of β^*, that is, $|\frac{\partial Q}{\partial \beta}(\tilde{\beta})| \leq \Theta_2$, $\Theta_2 > 0$, $\tilde{\beta} \in \mathcal{V}(\beta^*)$, where $\mathcal{V}(\beta^*)$ denotes a compact neighborhood of β^*.*

We can now state the following result.

Theorem 3.2. *Consider the system (3.1) for any $x_0 \in \mathbb{R}^n$, under Assumptions 3.1–3.6, with the feedback controller*

$$u = A^{-1}(\xi)(v_s(t, \xi) - b(\xi)) - A^{-1}(\xi)\frac{\partial V}{\partial \tilde{z}}^T k(t) \, d_2(e), \; k > 0,$$

$$v_s = (v_{s1}, \ldots, v_{sm})^T,$$

$$v_{si}(t, \xi) = \hat{y}_{id}^{(ri)} - K_{ri}^i(t)(y_i^{(ri-1)} - \hat{y}_{id}^{(ri-1)}) - \cdots - K_1^i(t)(y_i - \hat{y}_{id}),$$

$$i = 1, \ldots, m \tag{3.20}$$

where the state vector is reset following the resetting law $x(It_f) = x_0$, $I \in \{1, 2, \ldots\}$, *the desired trajectory vector is reset following* $\hat{y}_{id}(t) = y_{id}(t - (I - 1)t_f)$, $(I - 1)t_f \le t < It_f$, $I \in \{1, 2, \ldots\}$, *and* $K_j^i(t) \in \mathcal{K}$, $j = 1, \ldots, ri$, $i = 1, \ldots, m$ *are piecewise continuous gains switched at each iteration* I, $I \in \{1, 2, \ldots\}$, *following the update law*

$$
\begin{aligned}
K_j^i(t) &= K_{j-\text{nominal}}^i + \delta K_j^i(t), \\
\delta K_j^i(t) &= \delta \hat{K}_j^i((I - 1)t_f), \quad (I - 1)t_f \le t < It_f, \\
k(t) &= k_{\text{nominal}} + \delta k(t), \quad k_{\text{nominal}} > 0 \\
\delta k(t) &= \delta \hat{k}((I - 1)t_f), \quad (I - 1)t_f \le t < It_f, \quad I = 1, 2, 3, \ldots
\end{aligned}
\tag{3.21}
$$

where $\delta \hat{K}_j^i$, $\delta \hat{k}$ *are given by Eqs. (3.18), (3.19) and whereas the rest of the coefficients are defined similar to Theorem 3.1. Then, the obtained closed-loop impulsive time-dependent dynamic system (3.1), (3.18), (3.19), (3.20), and (3.21) is well posed. The tracking error* z *is uniformly bounded, and is steered at each iteration* I *toward the positive invariant set* $S_I = \{z \in \mathbb{R}^n | \ 1 - k_I \| \frac{\partial V}{\partial z} \| \ge 0\}$, $k_I = \beta_I(n + 1)$, *where* β_I *is the value of* β *at the Ith iteration. Furthermore,* $|Q(\beta(It_f)) - Q(\beta^*)| \le \Theta_2 \left(\frac{\Theta_1}{\omega_0} + \sqrt{\sum_{i=1,\ldots,m, \ j=1,\ldots,ri} a_{K_j^i}^2 + a_k^2} \right)$, $\Theta_1, \Theta_2 > 0$, *for* $I \to \infty$, *where* $\omega_0 = Max(\omega_{K_1^1}, \ldots, \omega_{K_{rm}^m}, \omega_k)$, *and* Q *satisfies Assumptions 3.7–3.9. Wherein, the vector* β *remains bounded over the iterations s.t.* $\|\beta((I + 1)t_f) - \beta(It_f)\| \le 0.5 t_f Max(a_{K_1^1}^2, \ldots, a_{K_{rm}^m}^2, a_k^2) \Theta_2 + t_f \omega_0 \sqrt{\sum_{i=1,\ldots,m \ j=1,\ldots,ri} a_{K_j^i}^2 + a_k^2}$, $I \in \{1, 2, \ldots\}$, *and satisfies asymptotically the bound* $\|\beta(It_f) - \beta^*\| \le \frac{\Theta_1}{\omega_0} + \sqrt{\sum_{i=1,\ldots,m \ j=1,\ldots,ri} a_{K_j^i}^2 + a_k^2}$, $\Theta_1 > 0$, *for* $I \to \infty$.

Proof. First, we discuss the well-posedness of the obtained closed-loop impulsive dynamical system. Indeed, the closed-loop system (3.1), (3.18), (3.19), (3.20), and (3.21), can be viewed as an impulsive time-dependent dynamical system (Haddad et al., 2006, pp. 18–19), with the trivial resetting law $\Delta x(t) = x_0$, for $t = It_f$, $I \in \{1, 2, \ldots\}$. In this case the resetting times given by It_f, $t_f > 0$ $I \in \{1, 2, \ldots\}$, are well defined and distinct. Furthermore, due to Assumption 3.1 and the smoothness of Eq. (3.20) (within each iteration), this impulsive dynamic system admits a unique solution in forward time, for any initial condition $x_0 \in \mathbb{R}^n$ (Haddad et al., 2006, p. 12). Finally, the fact that $t_f \ne 0$ excludes a Zeno behavior over a finite time interval (only a finite number of resets are possible over a finite time interval). Next, let us consider the system (3.1) with the initial condition x_0 (or equivalently the initial tracking error $z_0 = h(x_0) - y_d(0)$), under Assumptions 3.1–3.6, with the feedback controller (3.20), (3.21), for a given time interval $(I' - 1)t_f \leq t < I' t_f$, for any given $I' \in \{1, 2, \ldots\}$. Based on Theorem 3.1, there exists a Lyapunov function $V_{I'} = z^T P_{I'} z$, such that $\dot{V}_{I'} \leq \left(1 - k_{I'} \|\frac{\partial V_{I'}}{\partial \hat{z}}\|\right) \|\frac{\partial V_{I'}}{\partial \hat{z}}\| d_2$, where $P_{I'}$ is solution of the Lyapunov equation $P_{I'} \tilde{A}_{I'} + \tilde{A}_{I'}^T P_{I'} = -I$, wherein $\tilde{A}$ given by Eq. (3.10), with the gains for the iteration I'

$$K_{I'j}^i(t) = K_{j-\text{nominal}}^i + \delta K_j^i(t),$$

$$\delta K_j^i(t) = \delta \hat{K}_j^i((I' - 1)t_f), \quad (I' - 1)t_f \leq t < I' t_f,$$

$$k_{I'}(t) = k_{\text{nominal}} + \delta k(t), \quad k_{\text{nominal}} > 0,$$

$$\delta k(t) = \delta \hat{k}((I' - 1)t_f), \quad (I' - 1)t_f \leq t < I' t_f, \quad I' = 1, 2, 3, \ldots$$

(3.22)

which shows that z, starting from z_0 (for all the iterations $I' \in \{1, 2, \ldots\}$) is steered $\forall t \in [(I' - 1)t_f, I' t_f[$, toward the invariant set $S_{I'} = \left\{z \in \mathbb{R}^n | 1 - k_{I'} \|\frac{\partial V_{I'}}{\partial \hat{z}}\| \geq 0\right\}$, and this is valid for all $I' \in \{1, 2, 3, \ldots\}$. Furthermore, because at each switching point, that is, each new iteration I', we reset the system from the same bounded initial condition z_0, we can conclude about the uniform boundedness of the tracking error z. Next, we use the results presented in Rotea (2000) that characterize the learning cost function Q behavior along the iterations. First, based on Assumptions 3.7–3.9, the extremum-seeking (ES) nonlinear dynamics (3.18), (3.19), can be approximated by a linear averaged dynamic (using averaging approximation over time (Rotea, 2000, p 435, Definition 1)). Furthermore, $\exists \Theta_1$, ω^*, such that for all $\omega_0 = \text{Max}(\omega_{K_1^1}, \ldots, \omega_{K_{rm}^m}, \omega_k) > \omega^*$, the solution of the averaged model $\beta_{\text{aver}}(t)$ is locally close to the solution of the original ES dynamics, and satisfies (Rotea, 2000, p. 436)

$$\|\beta(t) - d(t) - \beta_{\text{aver}}(t)\| \leq \frac{\Theta_1}{\omega_0}, \quad \Theta_1 > 0, \; \forall t \geq 0,$$

with $d_{\text{vec}}(t) = (a_{K_1^1} \sin(\omega_{K_1^1} t - \frac{\pi}{2}), \ldots, a_{K_{rm}^m} \sin(\omega_{K_{rm}^m} t - \frac{\pi}{2}), a_k \sin(\omega_k t - \frac{\pi}{2}))^T$, and $\delta = (\delta_{K_1^1}, \ldots, \delta_{K_{rm}^m}, \delta_k)^T$. Moreover, because Q is analytic it can be approximated locally in $\mathcal{V}(\beta^*)$ with a quadratic function, for example, Taylor series up to second order. This, together with the proper choice of the dither signals as in Eq. (3.18), and the dither frequencies satisfying Eq. (3.19), allows to prove that β_{aver} satisfies (Rotea, 2000, p. 437)

$$\lim_{t\to\infty} \beta_{\mathrm{aver}}(t) = \beta^*,$$

which together with the previous inequality leads to

$$\|\beta(t) - \beta^*\| - \|d(t)\| \le \|\beta(t) - \beta^* - d(t)\| \le \frac{\Theta_1}{\omega_0}, \ \Theta_1 > 0, \ t \to \infty,$$

$$\Rightarrow \|\beta(t) - \beta^*\| \le \frac{\Theta_1}{\omega_0} + \|d(t)\|, \ t \to \infty.$$

This finally implies that

$$\|\beta(t) - \beta^*\| \le \frac{\Theta_1}{\omega_0} + \sqrt{\sum_{i=1,\dots,m,\, j=1,\dots,ri} a_{K_j^i}{}^2 + a_k{}^2}, \ \Theta_1 > 0, \ t \to \infty,$$

$$\Rightarrow \|\beta(It_f) - \beta^*\| \le \frac{\Theta_1}{\omega_0} + \sqrt{\sum_{i=1,\dots,m,\, j=1,\dots,ri} a_{K_j^i}{}^2 + a_k{}^2}, \ \Theta_1 > 0, \ I \to \infty.$$

Next, based on Assumption 3.9, the cost function is locally Lipschitz, with the Lipschitz constant $\max_{\beta \in \mathcal{V}(\beta^*)} \|\frac{\partial Q}{\partial \beta}\| = \Theta_2$, that is, $|Q(\beta_1) - Q(\beta_2)| \le \Theta_2 \|\beta_1 - \beta_2\|$, $\forall \beta_1, \beta_2 \in \mathcal{V}(\beta^*)$, which together with the previous inequality leads to

$$|Q(\beta(It_f)) - Q(\beta^*)| \le \Theta_2 \Big(\frac{\Theta_1}{\omega_0} + \sqrt{\sum_{i=1,\dots,m,\, j=1,\dots,ri} a_{K_j^i}{}^2 + a_k{}^2}\Big),$$

$$\Theta_1 > 0, \ \Theta_2 > 0, \ I \to \infty.$$

Finally, we show that the ES algorithm (3.18), (3.19) is a gradient-based algorithm, as follows: from Eq. (3.18), and if we denote $X = (x_{K_1^1}, \dots, x_{K_{rm}^m}, x_k)^T$, we can write

$$\dot{X} = \Big(a_{K_1^1}\omega_{K_1^1} \sin\Big(\omega_{K_1^1} t - \frac{\pi}{2}\Big), \dots, a_{K_{rm}^m}\omega_{K_{rm}^m} \sin\Big(\omega_{K_{rm}^m} t - \frac{\pi}{2}\Big),$$

$$a_k\omega_k \sin\Big(\omega_k t - \frac{\pi}{2}\Big)\Big)^T Q(\beta). \tag{3.23}$$

Based on Assumption 3.9, the cost function can be locally approximated with its first order Taylor development in $\mathcal{V}(\beta^*)$, which leads to

$$\dot{X} \simeq \tilde{d}_{\mathrm{vec}} \Big(Q(\tilde{\beta}) + \bar{d}_{\mathrm{vec}}^T \frac{\partial Q}{\partial \beta}(\tilde{\beta})\Big), \ \tilde{\beta} \in \mathcal{V}(\beta^*), \tag{3.24}$$

where $\tilde{d}_{\mathrm{vec}} = (a_{K_1^1}\omega_{K_1^1} \sin(\omega_{K_1^1} t - \frac{\pi}{2}), \dots, a_{K_{rm}^m}\omega_{K_{rm}^m} \sin(\omega_{K_{rm}^m} t - \frac{\pi}{2}),$ $a_k\omega_k \sin(\omega_k t - \frac{\pi}{2}))^T$, and $\bar{d}_{\mathrm{vec}} = (a_{K_1^1} \sin(\omega_{K_1^1} t + \frac{\pi}{2}), \dots, a_{K_{rm}^m} \sin(\omega_{K_{rm}^m} t + \frac{\pi}{2}), a_k \sin(\omega_k t + \frac{\pi}{2}))^T$.

Next, by integrating Eq. (3.24), over $[t,\ t + t_f]$ and neglecting the terms inversely proportional to the high frequencies, that is, terms on $\frac{1}{\omega_i}$'s (high frequencies filtered by the integral operator), we obtain

$$X(t + t_f) - X(t) \simeq -t_f R \frac{\partial Q}{\partial \beta}(\tilde{\beta}), \qquad (3.25)$$

with $R = 0.5\ \text{diag}\{\omega_{K_1^1} a_{K_1^1}{}^2, \dots, \omega_{K_{rm}^m} a_{K_{rm}^m}{}^2, \omega_k a_k^2\}$.

Next, from Eqs. (3.14), (3.18), we can write $\|\beta(t + t_f) - \beta(t)\| \le \|X(t + t_f) - X(t)\| + \|\bar{d}_{\text{vec}}(t + t_f) - \bar{d}_{\text{vec}}(t)\|$, which together with Eq. (3.25), with the bound $\|\frac{\bar{d}_{\text{vec}}(t + t_f) - \bar{d}_{\text{vec}}(t)}{t_f}\| \le \|\dot{\bar{d}}_{\text{vec}}\| \le \omega_0 \sqrt{\sum_{i=1,\dots,m\ j=1,\dots,ri} a_{K_j^i}{}^2 + a_k^2}$, and Assumption 3.9, leads to the inequality

$$\|\beta((I + 1)t_f) - \beta(It_f)\| \le 0.5 t_f \text{Max}(\omega_{K_1^1} a_{K_1^1}{}^2, \dots, \omega_{K_{rm}^m} a_{K_{rm}^m}{}^2, \omega_k a_k^2)\Theta_2$$

$$+ t_f \omega_0 \sqrt{\sum_{i=1,\dots,m\ j=1,\dots,ri} a_{K_j^i}{}^2 + a_k^2}, \ I \in \{1, 2, \dots\}.$$

Remark 3.4.

- *The learning procedure summarized in Theorem 3.2 should be interpreted as an iterative auto-tuning of the gains, by repeating the same task I iterations. This approach can be seen as an extension to nonlinear systems of the IFT algorithms (e.g., Hjalmarsson (2002)).*

- *The asymptotic convergence bounds presented in Theorem 3.2 are correlated to the choice of the first-order multiparametric extremum seeking (3.18); however, these bounds can be easily changed by using other MES algorithms, see Noase et al. (2011) and Scheinker (2013). This is due to the modular design of the controller (3.20), (3.21), which uses the robust part to ensure boundedness of the tracking error dynamics, and the learning part to optimize the cost function Q.*

- *In Theorem 3.2, we show that in each iteration I, the tracking error vector z is steered toward the invariant set S_I. However, due to the finite time-interval length t_f of each iteration, we cannot guarantee that the vector z enters S_I in each iteration (unless we are in the trivial case where $z_0 \in S_I$). All we guarantee is that the vector norm $\|z\|$ starts from a bounded value $\|z_0\|$ and remains bounded during the iterations with an upper-bound which can be estimated as function of $\|z_0\|$ by using the bounds of the quadratic Lyapunov functions V_I, $I = 1, 2, \dots$, that is, a uniform boundedness result (Haddad et al., 2006, p. 6, Definition 2.12).*

In the next section we propose to illustrate this approach on two mechatronics systems.

3.5 MECHATRONICS EXAMPLES

3.5.1 Electromagnetic Actuators

We apply here the method presented previously to the case of electromagnetic actuators. This system requires accurate control of a moving armature between two desired positions. The main objective, known as soft landing of the moving armature, is to ensure small contact velocity between the moving armature and the fixed parts of the actuator. This motion is usually of iterative nature because the actuator has to iteratively open and close to achieve a desired cyclic motion of a mechanical part attached to the actuator, for example, engine-valve systems in automotive applications.

3.5.1.1 System Modeling

Following Wang et al. (2000) and Peterson and Stefanopoulou (2004), we consider the following nonlinear model for electromagnetic actuators:

$$
\begin{aligned}
m\frac{d^2 x_a}{dt^2} &= k(x_0 - x_a) - \eta\frac{dx_a}{dt} - \frac{ai^2}{2(b + x_a)^2}, \\
u &= Ri + \frac{a}{b + x_a}\frac{di}{dt} - \frac{ai}{(b + x_a)^2}\frac{dx_a}{dt}, \quad 0 \leq x_a \leq x_f,
\end{aligned}
\tag{3.26}
$$

where x_a represents the armature position physically constrained between the initial position of the armature 0, and the maximal position of the armature x_f, $\frac{dx_a}{dt}$ represents the armature velocity, m is the armature mass, k the spring constant, x_0 the initial spring length, η the damping coefficient (assumed to be constant), $\frac{ai^2}{2(b + x_a)^2}$ represents the electromagnetic force (EMF) generated by the coil, a, b are two constant parameters of the coil, R the resistance of the coil, $L = \frac{a}{b + x_a}$ the coil inductance, and $\frac{ai}{(b + x_a)^2}\frac{dx_a}{dt}$ represents the back EMF. Finally, i denotes the coil current, $\frac{di}{dt}$ its time derivative, and u represents the control voltage applied to the coil. In this model, we do not consider the saturation region of the flux linkage in the magnetic field generated by the coil because we assume a current and armature motion ranges within the linear region of the flux.

3.5.1.2 Robust Controller

In this section we first design a nonlinear robust control based on Theorem 3.1. Following Assumption 3.4 we define x_{ref} a desired armature position trajectory, s.t., x_{ref} is a smooth (at least C^2) function satisfying the initial/final constraints $x_{ref}(0) = 0$, $x_{ref}(t_f) = x_f$, $\dot{x}_{ref}(0) = 0$, $\dot{x}_{ref}(t_f) = 0$, where t_f is a desired finite motion time, and x_f is a desired final position.

We consider the dynamical system (3.26) with bounded parametric uncertainties on the spring coefficient δk, with $|\delta k| \leq \delta k_{max}$, and the damping coefficient $\delta \eta$, with $|\delta \eta| \leq \delta \eta_{max}$, such that $k = k_{nominal} + \delta k$, $\eta = \eta_{nominal} + \delta \eta$, where $k_{nominal}$ and $\eta_{nominal}$ are the nominal values of the spring stiffness and the damping coefficient, respectively. If we consider the state vector $x = (x_a, \dot{x}_a, i)'$, and the controlled output x_a, the uncertain model of electromagnetic actuators can be written in the form of Eq. (3.1), as

$$
\dot{x} = \begin{pmatrix} \dot{x}_a \\ \ddot{x}_a \\ \dot{i} \end{pmatrix} = \begin{pmatrix} x_2 \\ \frac{k_{nominal}}{m}(x_0 - x_1) - \frac{\eta_{nominal}}{m}x_2 - \frac{ax_3^2}{2(b+x_1)^2} \\ -\frac{R(b+x_1)}{a}x_3 + \frac{x_3 x_2}{b+x_1} \end{pmatrix}
$$

$$
+ \begin{pmatrix} 0 \\ \frac{\delta k}{m}(x_0 - x_1) + \frac{\delta \eta}{m}x_2 \\ 0 \end{pmatrix} + \begin{pmatrix} 0 \\ 0 \\ \frac{b+x_1}{a} \end{pmatrix} u, \quad (3.27)
$$

$$
y = x_1.
$$

Assumption 3.1 is clearly satisfied over a nonempty bounded states set X. As for Assumption 3.2, it is straightforward to check that if we compute the third time-derivative of the output x_a, the control variable u appears in a nonsingular expression, which implies that $r = n = 3$. Assumption 3.3 is also satisfied because $\|\Delta f(x)\| \leq \frac{\delta k_{max}}{m}|x_0 - x_1| + \frac{\delta \eta_{max}}{m}|x_2|$.

Next, following the input-output linearization method, we can write

$$
y^{(3)} = x_a^{(3)} = -\frac{k_{nominal}}{m}\dot{x}_a - \frac{\eta_{nominal}}{m}\ddot{x}_a + \frac{Ri^2}{(b+x_a)m} - \frac{\delta k}{m}\dot{x}_a
$$

$$
- \frac{\delta \eta}{m}\ddot{x}_a - \frac{i}{m(b+x_a)}u, \quad (3.28)
$$

which is of the form of Eq. (3.4), with $A = -\frac{i}{m(b+x_a)}$, $b = -\frac{k_{nominal}}{m}\dot{x}_a - \frac{\eta_{nominal}}{m}\ddot{x}_a + \frac{Ri^2}{(b+x_a)m}$, and the additive uncertainty term $\Delta b = -\frac{\delta k}{m}\dot{x}_a - \frac{\delta \eta}{m}\ddot{x}_a$, such that $|\Delta b| \leq \frac{\delta k_{max}}{m}|\dot{x}_a| + \frac{\delta \eta_{max}}{m}|\ddot{x}_a| = d_2(x_a, \dot{x}_a)$.

Let us define the tracking error vector $\mathbf{z} := (z_1, z_2, z_3)^T = (x_a - x_{ref}, \dot{x}_a - \dot{x}_{ref}, \ddot{x}_a - \ddot{x}_{ref})^T$, where $\dot{x}_{ref} = \frac{dx_{ref}(t)}{dt}$ and $\ddot{x}_{ref} = \frac{d^2 x_{ref}(t)}{dt^2}$. Then, using Theorem 3.1, we can write the following robust controller

$$u = -\frac{m(b + x_a)}{i}\left(v_s + \frac{k_{\text{nominal}}}{m}\dot{x}_a + \frac{\eta_{\text{nominal}}}{m}\ddot{x}_a - \frac{Ri^2}{(b + x_a)m}\right)$$
$$+ \frac{m(b + x_a)}{i}\frac{\partial V}{\partial z_3}k\left(\frac{\delta k_{\max}}{m}|\dot{x}_a| + \frac{\delta \eta_{\max}}{m}|\ddot{x}_a|\right),$$
$$v_s = x_{\text{ref}}^{(3)}(t) + K_3(x_a^{(2)} - x_{\text{ref}}^{(2)}(t)) + K_2(x_a^{(1)} - x_{\text{ref}}^{(1)}(t)) + K_1(x_a - x_{\text{ref}}(t)),$$
$$k > 0, \; K_i < 0, i = 1, 2, 3$$

$$(3.29)$$

where $V = z^T P z$, $P > 0$ is a solution of the equation $P\tilde{A} + \tilde{A}^T P = -I$, with

$$\tilde{A} = \begin{pmatrix} 0 & 1 & 0 \\ 0 & 0 & 1 \\ K_1 & K_2 & K_3 \end{pmatrix}, \tag{3.30}$$

where K_1, K_2, and K_3 are chosen such that $\tilde{A}$ is Hurwitz.

Remark 3.5. *Regarding Assumption 3.6, about the existence of a nonempty set of gains $\mathcal{K}$ such that $\tilde{A}$ is Hurwitz, we can easily characterize $\mathcal{K}$ in this case. Indeed, if we want to place the eigenvalues of $\tilde{A}$ at the values s_1, s_2, s_3, such that $s_{1_{\min}} \le s_1 \le s_{1_{\max}}$, $s_{1_{\min}} < s_{1_{\max}} < 0$, $s_{2_{\min}} \le s_2 \le s_{2_{\max}}$, $s_{2_{\min}} < s_{2_{\max}} < 0$, and $s_{3_{\min}} \le s_3 \le s_{3_{\max}}$, $s_{3_{\min}} < s_{3_{\max}} < 0$, by direct match of the coefficients of the characteristic polynomial $s^3 - K_3 s^2 - K_2 s - K_1 = 0$, with the desired characteristic polynomial $\prod_{i=1}^{i=3}(s - s_i) = 0$, we can write*

$$K_1 = \prod_{i=1}^{i=3} s_i,$$
$$K_2 = -\sum_{i,j \in \{1,2,3\}, i \ne j} s_i s_j, \tag{3.31}$$
$$K_3 = \sum_{i=1}^{i=3} s_i,$$

which allows us to write the set $\mathcal{K}$ as

$$\mathcal{K} = \{(K_1, K_2, K_3) | s_{i_{\min}} s_{j_{\min}} s_{k_{\max}} \le K_1 \le s_{i_{\max}} s_{j_{\max}} s_{k_{\min}},$$
$$i \ne j \ne k, i, j, k \in \{1, 2, 3\},$$
$$-\sum_{i,j \in \{1,2,3\}, i \ne j} s_{i_{\min}} s_{j_{\min}} \le K_2 \le -\sum_{i,j \in \{1,2,3\}, i \ne j} s_{i_{\max}} s_{j_{\max}},$$
$$\sum_{i=1}^{i=3} s_{i_{\min}} \le K_3 \le \sum_{i=1}^{i=3} s_{i_{\max}}\}.$$

3.5.1.3 Learning-Based Auto-Tuning of the Controller Gains

We use now the results of Theorem 3.2 to iteratively auto-tune the gains of the feedback controller (3.29).

Considering a cyclic behavior of the actuator, such that each iteration happens over a time interval of length t_f, and following Eq. (3.16) we define the cost function as

$$Q(z(\beta)) = C_1 z_1 (It_f)^2 + C_2 z_2 (It_f)^2, \tag{3.32}$$

where $I = 1, 2, 3, \ldots$ is the number of iterations, C_1, $C_2 > 0$, and $\beta = (\delta K_1,\ \delta K_2,\ \delta K_3,\ \delta k)^T$, such as the feedback gains write as

$$\begin{aligned}
K_1 &= K_{1_{\text{nominal}}} + \delta K_1, \\
K_2 &= K_{2_{\text{nominal}}} + \delta K_2, \\
K_3 &= K_{3_{\text{nominal}}} + \delta K_3, \\
k &= k_{\text{nominal}} + \delta k,
\end{aligned} \tag{3.33}$$

where $K_{1_{\text{nominal}}}$, $K_{2_{\text{nominal}}}$, $K_{3_{\text{nominal}}}$, and k_{nominal} are the nominal initial values of the feedback gains in Eq. (3.29).

Following Eq. (3.18), (3.19), (3.21) the variations of the estimated gains are given by

$$\dot{x}_{K_1} = a_{K_1} \sin\left(\omega_1 t - \frac{\pi}{2}\right) Q(z(\beta)),$$

$$\delta \hat{K}_1(t) = x_{K_1}(t) + a_{K_1} \sin\left(\omega_1 t + \frac{\pi}{2}\right),$$

$$\dot{x}_{K_2} = a_{K_2} \sin\left(\omega_2 t - \frac{\pi}{2}\right) Q(z(\beta)),$$

$$\delta \hat{K}_2(t) = x_{K_2}(t) + a_{K_2} \sin\left(\omega_2 t + \frac{\pi}{2}\right),$$

$$\dot{x}_{K_3} = a_{K_3} \sin\left(\omega_3 t - \frac{\pi}{2}\right) Q(z(\beta)),$$

$$\delta \hat{K}_3(t) = x_{K_3}(t) + a_{K_3} \sin\left(\omega_3 t + \frac{\pi}{2}\right),$$

$$\dot{x}_k = a_k \sin\left(\omega_4 t - \frac{\pi}{2}\right) Q(z(\beta)),$$

$$\delta \hat{k}(t) = x_k(t) + a_k \sin\left(\omega_4 t + \frac{\pi}{2}\right),$$

$$\delta K_j(t) = \delta \hat{K}_j((I-1)t_f),\ (I-1)t_f \leq t < It_f,\ j \in \{1, 2, 3\},\ I = 1, 2, 3, \ldots$$

$$\delta k(t) = \delta \hat{k}((I-1)t_f),\ (I-1)t_f \leq t < It_f,\ I = 1, 2, 3, \ldots$$

$$\tag{3.34}$$

Table 3.1 Numerical values of the mechanical parameters for the electromagnetic example

Parameter	Value
m	0.3 kg
R	6.5 Ω
η	8 kg/s
x_0	8 mm
k	160 N/mm
a	15×10^{-6} Nm^2/A^2
b	4.5×10^{-5} mm

where a_{K_1}, a_{K_2}, a_{K_3}, and a_k are positive, and $\omega_p + \omega_q \neq \omega_r$, $p, q, r \in \{1, 2, 3, 4\}$, for $p \neq q \neq r$.

3.5.1.4 Simulation Results

We apply here the proposed approach to the electromagnetic actuator with the physical constants reported in Table 3.1. For simplicity, the desired trajectory has been selected as the fifth–order polynomial $x_{ref}(t) = \sum_{i=0}^{5} a_i(t/t_f)^i$, where the a_i's have been computed to satisfy the boundary constraints $x_{ref}(0) = 0, x_{ref}(t_f) = x_f, \dot{x}_{ref}(0) = \dot{x}_{ref}(t_f) = 0, \ddot{x}_{ref}(0) = \ddot{x}_{ref}(t_f) = 0$, with $t_f = 1$ s, $x_f = 0.5$ mm.

Furthermore, to make the simulation case more challenging we assume an initial error both on the position and the velocity $z_1(0) = 0.01$ mm, $z_2(0) = 0.1$ mm/s. Note that these values may seem small, but for this type of actuator it is usually the case that the armature starts from a predefined static position constrained mechanically, so we know that the initial velocity is zero and we know in advance very precisely the initial position of the armature. However, we want to show the performances of the controller on some challenging cases. We first select the nominal feedback gains $K_1 = -800$, $K_2 = -200$, $K_3 = -40$, $k = 1$, satisfying Assumption 3.5.

In the first test we compare the performances of the robust controller (3.29) with fixed nominal gains, to the learning controller (3.29), (3.33), and (3.34), which was implemented with the cost function (3.32) where $C_1 = 500$, $C_2 = 500$, and the learning frequencies for each feedback gain are $\omega_1 = 7.5$ rad/s, $\omega_2 = 5.3$ rad/s, $\omega_3 = 5.1$ rad/s, and $\omega_4 = 6.1$ rad/s. It is well known in the MES community that the learning convergence rate is related to the choice of the coefficients $a_{K_i}, i = 1, 2, 3, a_k$, see Tan et al. (2008). First, we will implement the base line MES algorithm as

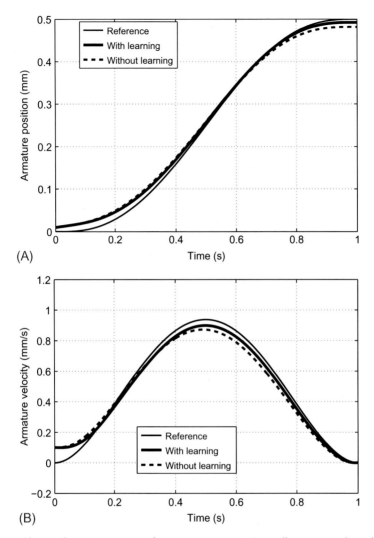

Fig. 3.1 Obtained outputs versus reference trajectory—Controller (3.29) without learning (*dashed line*), with learning (*bold line*). (A) Obtained armature position versus reference trajectory. (B) Obtained armature velocity versus reference trajectory.

presented in Section 3.5.1.3, with constant MES search amplitudes a_{K_i}. We select the following amplitudes $a_{K_1} = 100$, $a_{K_2} = 5$, $a_{K_3} = 1$, and $a_k = 1$. Fig. 3.1A and B displays the performance of the position and the velocity tracking, with and without the learning algorithm. We see clearly the effect of the learning algorithm that makes the armature position and velocity closer to the desired trajectories. However, even though the actual trajectories are getting closer to the desired ones, the tracking is still not exact. There are two main reasons for this: First, as we will see later when we analyze the cost function graphs, we have stopped the learning iteration too early, in the sense that the cost function is still decreasing and did not reach a local minimum yet. The reason we stopped the learning iterations is that we wanted to compare this learning algorithm (with constant search amplitude) to other algorithms (with time-varying search amplitudes), for the same number of iterations. Second, the choice of the cost function in this first set of tests is given by Eq. (3.32), which does not seek an exact trajectory tracking, but rather a soft landing, that is, making the actual trajectories close to the desired trajectories at the impact time, which is what the learning algorithm ends up doing. This point is very important because this shows us the importance of the cost function choice in this type of learning-based tuning algorithm. Indeed, the cost-function needs to be selected to represent the desired performance of the controller. We will see later in this section that with a different choice of the cost function we can obtain optimal gains that target other control performances, for instance, trajectory tracking over the whole time interval. We report in Fig. 3.2 the cost function value as function of the learning iterations. We see a clear decrease of the cost function; however, the decease rate is rather

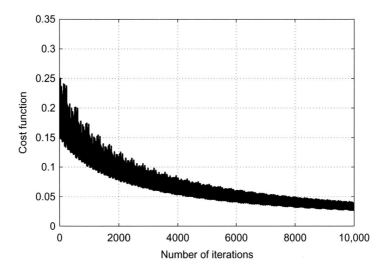

Fig. 3.2 Cost function—Test 1.

slow. We will see that using time-varying search amplitudes can accelerate the decrease rate. Next, we report the learned feedback gains in Fig. 3.3. They also show a trend of slow convergence, with large oscillations around the mean value. The excursion of these oscillations can be easily tuned by the proper selection of the learning coefficients a_{K_i}, $i = 1, 2, 3, 4$,

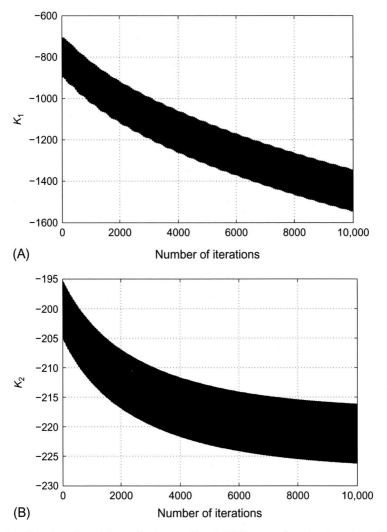

(A)

Number of iterations

(B)

Number of iterations

Fig. 3.3 Gains learning—Controller (3.29)—Test 1. (A) K_1 versus learning iterations. (B) K_2 versus learning iterations.

(Continued)

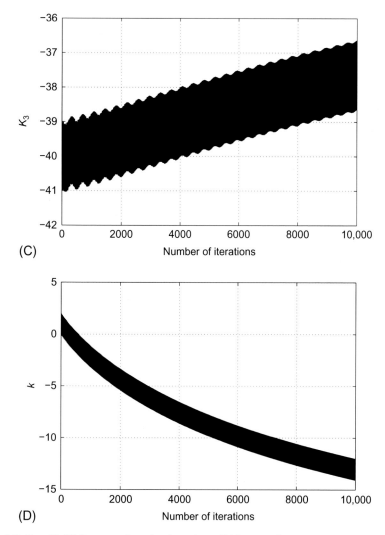

Fig. 3.3, Cont'd (C) K_3 versus learning iterations. (D) k versus learning iterations.

as we will see in the next set of tests, when we use time-varying search amplitudes a_{K_i}.

Indeed, in this second test we use varying values for the coefficients a_{K_i}. It is well known, see Moase et al. (2009), that choosing varying coefficients, which start with a high value to accelerate the search initially and then are tuned down when the cost function becomes smaller, accelerates the learning and achieves a convergence to a tighter neighborhood of the local optimum (due to the decrease of the dither amplitudes). To implement this idea in the following test, we simply use piecewise constant coefficients as follows: $a_{K_1} = 1000$, $a_{K_2} = 500$, $a_{K_3} = 100$, $a_k = 1$ initially, and then tune them down to $a_{K_1} = 1000Q(1)/4$, $a_{K_2} = 500Q(1)/4$, $a_{K_3} = 100(1)/4$, $a_k = Q(1)/4$, when $Q \leq Q(1)/2$, and then to $a_{K_1} = 1000Q(1)/5$, $a_{K_2} = 500Q(1)/5$, $a_{K_3} = 100Q(1)/5$, $a_k = Q(1)/5$, when $Q \leq Q(1)/3$, where $Q(1)$ denotes the value of the cost function at the first iteration. We report in Fig. 3.4 the cost function value along the learning iterations. We see a clear decrease of the cost function, which is noticeably faster than the decrease of the cost function obtained with constant coefficients shown in Fig. 3.2. We see that the cost function drops below half of the initial value within 100 iterations. However, due to the abrupt switch of the search amplitudes a_{Ki}'s we notice few abrupt jumps in the cost function. We will see in Test 3 that we can smoothen the learning convergence further by selecting a different way to change the search amplitudes. The learned feedback gains are reported in Fig. 3.5. They also show a rate of convergence which is much faster than the rate of convergence of Test 1. We also notice that the gains in Test 2 do not converge to the same values as in Test 1; this is expected due to the local nature of the MES search algorithm. However, even if we do not reach global optimal gains, we still have a clear improvement of the closed-loop performance as proven by the decrease of the learning cost function. One can see in Fig. 3.5 that the gains converge toward a mean value, but with remaining large excursions around the mean. This is due to the fact that we have chosen piecewise constant value for the amplitudes a_{K_i}'s, which leads to constant excursions even after the decrease of the cost function below an acceptable threshold value. One straightforward way to remove this final excursion is to simply keep the gains constant, that is, stop the learning, when the cost function drops below a desired threshold value. Another method to obtain a smoother learning with smaller residual oscillations of the gains is presented in Test 3.

In this set of simulations, we propose to fine-tune the amplitudes of the search amplitudes a_{K_i}'s by relating them to the amplitude of the cost function; we will refer to this MES algorithm as a dynamic MES algorithm compared to the static MES with constant a_{K_i}'s used in Test 1. To obtain a smooth change of the search amplitudes, we propose to slightly change the original MES algorithm (3.45) as follows:

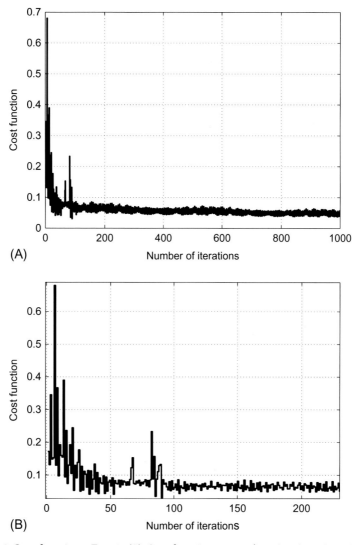

Fig. 3.4 Cost function—Test 2. (A) Cost function versus learning iterations. (B) Cost function versus learning iterations—Zoom.

$$\dot{x}_{K_1} = a_{K_1} \sin\left(\omega_1 t - \frac{\pi}{2}\right) Q(z(\beta)),$$

$$\delta \hat{K}_1(t) = x_{K_1}(t) + a_{K_1} Q(z(\beta)) \sin\left(\omega_1 t + \frac{\pi}{2}\right),$$

$$\dot{x}_{K_2} = a_{K_2} \sin\left(\omega_2 t - \frac{\pi}{2}\right) Q(z(\beta)),$$

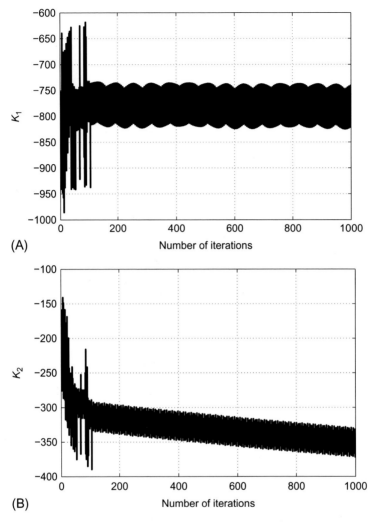

Fig. 3.5 Gains learning—Controller (3.29)—Test 2. (A) K_1 versus learning iterations. (B) K_2 versus learning iterations.

(Continued)

$$\delta \hat{K}_2(t) = x_{K_2}(t) + a_{K_2} Q(z(\beta)) \sin\left(\omega_2 t + \frac{\pi}{2}\right),$$

$$\dot{x}_{K_3} = a_{K_3} \sin\left(\omega_3 t - \frac{\pi}{2}\right) Q(z(\beta)),$$

$$\delta \hat{K}_3(t) = x_{K_3}(t) + a_{K_3} Q(z(\beta)) \sin\left(\omega_3 t + \frac{\pi}{2}\right),$$

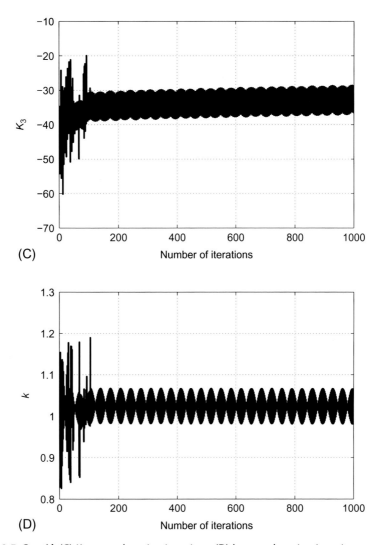

Fig. 3.5, Cont'd (C) K_3 versus learning iterations. (D) k versus learning iterations.

$$\dot{x}_k = a_k \sin\left(\omega_4 t - \frac{\pi}{2}\right) Q(z(\beta)),$$

$$\delta \hat{k}(t) = x_k(t) + a_k Q(z(\beta)) \sin\left(\omega_4 t + \frac{\pi}{2}\right),$$

$$\delta K_j(t) = \delta \hat{K}_j((I-1)t_f), \ (I-1)t_f \le t < It_f, \ j \in \{1,2,3\}, \ I = 1,2,3,\ldots$$

$$\delta k(t) = \delta \hat{k}((I-1)t_f), \ (I-1)t_f \le t < It_f, \ I = 1,2,3,\ldots$$

$$(3.35)$$

In this case, the search amplitudes are function of the cost-function value and are expected to decrease together with the decrease of Q. We implemented the feedback controller (3.29), (3.33), and (3.35), with the coefficients $a_{K_1} = 1000$, $a_{K_2} = 500$, $a_{K_3} = 100$, and $a_k = 1$. The obtained cost function is reported in Fig. 3.6, where we can see that the dynamic

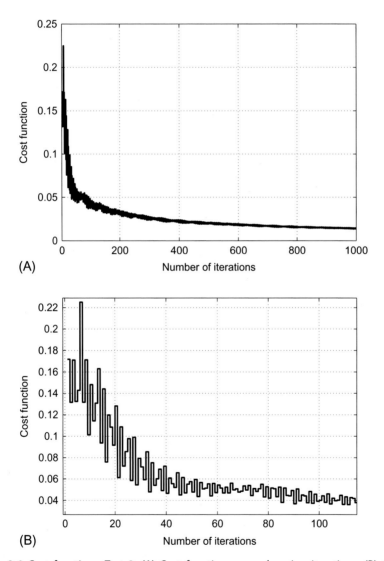

(A)

(B)

Fig. 3.6 Cost function—Test 3. (A) Cost function versus learning iterations. (B) Cost function versus learning iterations—Zoom.

MES results in a much smoother decrease of the cost function compared to Test 2, and faster decrease compared to Test 1, because in this case the cost function decreases below half of its initial value within 20 iterations. The corresponding feedback gains learning trend is shown in Fig. 3.7.

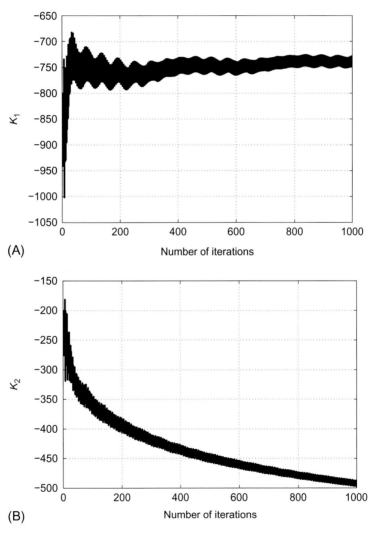

Fig. 3.7 Gains learning—Controller (3.29)—Test 3. (A) K_1 versus learning iterations. (B) K_2 versus learning iterations.

(Continued)

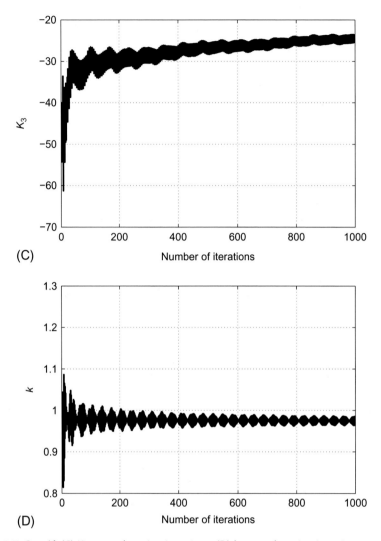

Fig. 3.7, Cont'd (C) K_3 versus learning iterations. (D) k versus learning iterations.

If we compare these gains to the one obtained in Test 2 (shown in Fig. 3.5), we can see that the transient part of the learning (the first iterations) do not show abrupt big jumps like in Test 2; this is due to the smooth change of the search amplitude as function of the cost function. Furthermore, the excursions amplitudes of the gains are quickly decreasing because they are directly proportional to the cost value. Here again, we could just stop the learning when the cost function has decreased below a predefined threshold, which will happen faster than in the case of Test 1 and Test 2.

Our conclusion is that in these tests, where the control goal is a soft landing of the armature as indicated by the cost function (3.32), the dynamic MES implies faster and smoother learning of the gains than the static MES (Test 1), or the quasistatic MES (Test 2).

Next, in Test 4, we will evaluate the learning algorithm when dealing with another control objective. In the next simulations we reformulate the cost function to model a tracking control objective. We consider the following tracking cost function:

$$Q(z(\beta)) = \int_{(I-1)t_f}^{It_f} z_1^T(t) C_1 z_1(t) dt + \int_{(I-1)t_f}^{It_f} z_2^T(t) C_2 z_2(t) dt, \quad (3.36)$$

where $C_1 > 0$, $C_2 > 0$. Choosing this cost function means that in this test we focus on tuning the feedback gains in such a way to make the armature position and velocity follow some desired trajectories. Because we saw in the previous tests that the dynamic MES gives the best convergence performances, we only show here the performance of the dynamic MES given by Eq. (3.35). We see in Fig. 3.8 that due to the new formulation of the cost function, the armature position and velocity follows better the desired reference trajectory than in the previous tests (see Fig. 3.1). The corresponding decreasing cost function is shown in Fig. 3.9. Finally, the gains' learning profile is shown in Fig. 3.10.

Here again to avoid the gains' oscillation and long drift, one could simply switch off the learning when the cost function reaches a low value that is considered to be a good-enough performance.

3.5.2 Two-Link Rigid Manipulators

We consider now the problem of output trajectory tracking for robot manipulators. To simplify the presentation we focus on the case of two–link rigid arms; however, the same result can be readily extended to the case of n links. The objective of the controller is to make the robot joints' angle track some desired angular time-trajectories, where the feedback gains are auto–tuned online.

3.5.2.1 System Modeling

It is well established in the robotics literature that the dynamics for rigid links manipulators is given by, see Spong (1992)

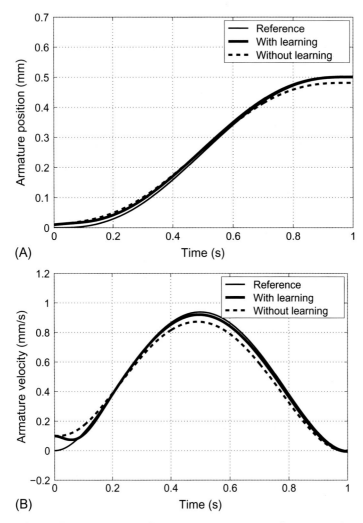

Fig. 3.8 Obtained outputs versus reference trajectory—Controller (3.29) without learning (*dashed line*), with learning (*bold line*). (A) Obtained armature position versus reference trajectory— (B) Obtained armature velocity versus reference trajectory.

$$H(q)\ddot{q} + C(q,\dot{q})\dot{q} + D\dot{q} + G(q) = \tau, \tag{3.37}$$

where $q \triangleq (q_1, q_2)^T$ denotes the two joints' angles and $\tau \triangleq (\tau_1, \tau_2)^T$ denotes the two joints' torques. The matrix H is assumed to be nonsingular, and is given by

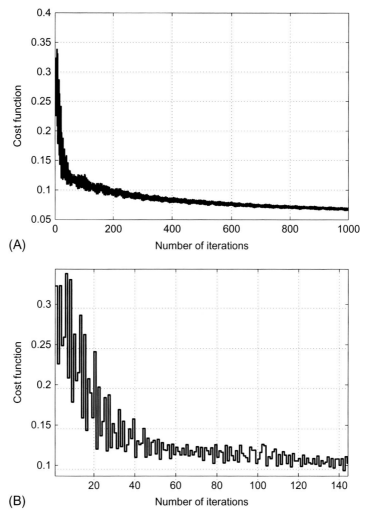

Fig. 3.9 Cost function—Test 4. (A) Cost function versus learning iterations. (B) Cost function versus learning iterations—Zoom.

$$H \triangleq \begin{pmatrix} H_{11} & H_{12} \\ H_{21} & H_{22} \end{pmatrix},$$

where

$$H_{11} = m_1 \ell_{c_1}^2 + I_1 + m_2 [\ell_1^2 + \ell_{c_2}^2 + 2\ell_1 \ell_{c_2} \cos(q_2)] + I_2,$$

$$H_{12} = m_2 \ell_1 \ell_{c_2} \cos(q_2) + m_2 \ell_{c_2}^2 + I_2,$$

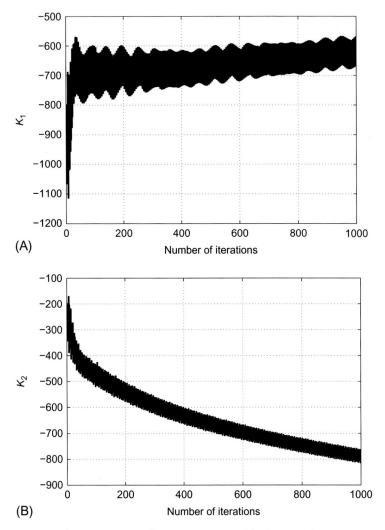

Fig. 3.10 Gains learning—Controller (3.29)—Test 4. (A) K_1 versus learning iterations. (B) K_2 versus learning iterations.

(Continued)

$$H_{21} = H_{12},$$
$$H_{22} = m_2 \ell_{c_2}^2 + I_2. \tag{3.38}$$

The matrix $C(q, \dot{q})$ is given by

$$C(q, \dot{q}) \triangleq \begin{pmatrix} -h\dot{q}_2 & -h\dot{q}_1 - h\dot{q}_2 \\ h\dot{q}_1 & 0 \end{pmatrix},$$

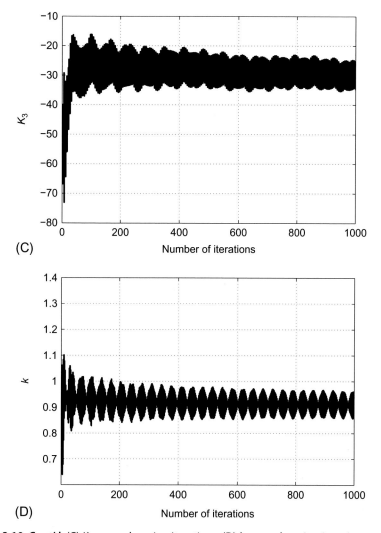

Fig. 3.10, Cont'd (C) K_3 versus learning iterations. (D) k versus learning iterations.

where $h = m_2 \ell_1 \ell_{c_2} \sin(q_2)$. The damping matrix D is given by

$$D \triangleq \begin{pmatrix} \eta_1 & 0 \\ 0 & \eta_2 \end{pmatrix}, \eta_1 > 0, \ \eta_2 > 0.$$

The vector $G = (G_1, G_2)^T$ is given by

$$\begin{aligned} G_1 &= m_1 \ell_{c_1} g \cos(q_1) + m_2 g[\ell_2 \cos(q_1 + q_2) + \ell_1 \cos(q_1)], \\ G_2 &= m_2 \ell_{c_2} g \cos(q_1 + q_2), \end{aligned} \tag{3.39}$$

where l_1 and l_2 are the lengths of the first and second link, respectively; m_1 and m_2 are the masses of the first and second link, respectively; I_1 is the moment of inertia of the first link and I_2 the moment of inertia of the second link; η_1 and η_2 are the damping coefficients for the first and second joint, respectively; and g denotes the earth gravitational constant.

3.5.2.2 Robust Controller

We consider the uncertain version of the model (3.37)

$$H(q)\ddot{q} + C(q,\dot{q})\dot{q} + (D + \delta D)\dot{q} + G(q) + d(t) = \tau, \qquad (3.40)$$

with bounded uncertainties of the damping matrix $\delta D = \mathrm{diag}(\delta\eta_1, \delta\eta_2)$, $\delta\eta_1 \leq \delta\eta_{1\max}$, $\delta\eta_2 \leq \delta\eta_{2\max}$, and an additive bounded disturbance $d(t) = (d_1, d_2)^T$, s.t. $\|d(t)\| \leq d_{\max}$, $\forall t$.

It is easy to verify Assumption 3.1 in this case. To check Assumption 3.2, we compute successive derivatives of y until the input vector τ appears, which in this case corresponds to the relative degree vector $r = (2,2)^T$.

Next, it is straightforward to check that the model (3.37) is of the form of Eq. (3.2), with $A = H^{-1}$, $b = H^{-1}(C\dot{q} + D\dot{q} + G)$. Then, using Theorem 3.1, we write the following robust controller:

$$\tau = C\dot{q} + D\dot{q} + G(q) + Hv_s$$
$$+ \left(\frac{\partial V}{\partial \dot{q}_1}, \frac{\partial V}{\partial \dot{q}_2}\right)^T k\|H^{-1}\|_F \left(\delta\eta_{\max}\sqrt{\dot{q}_1^2 + \dot{q}_2^2} + d_{\max}\right),$$
$$v_s = y_{\text{ref}}^{(2)} - K_D(y^{(1)} - y_{\text{ref}}^{(1)}) - K_P(y - y_{\text{ref}}),$$
$$k > 0, \ \delta\eta_{\max} = Max(\delta\eta_{1\max}, \delta\eta_{2\max}),$$
$$K_D = \mathrm{diag}(k_{d1}, k_{d2}) > 0, \ K_P = \mathrm{diag}(k_{p1}, k_{p2}) > 0,$$

$$(3.41)$$

where $y_{\text{ref}}(t) = (q_{1\text{ref}}(t), q_{2\text{ref}}(t))^T$ is a desired output trajectory, which is at least C^2. The Lyapunov function is defined as $V = z^T P z$, where $z = (z_1, z_2, z_3, z_4)^T = (q_1 - q_{1\text{ref}}(t), \dot{q}_1 - \dot{q}_{1\text{ref}}(t), q_2 - q_{2\text{ref}}(t), \dot{q}_2 - \dot{q}_{2\text{ref}}(t))^T$, and $P > 0$ is the solution of the Lyapunov equation $P\tilde{A} + \tilde{A}^T P = -I$, with

$$\tilde{A} = \begin{pmatrix} 0 & 1 & 0 & 0 \\ -k_{p1} & -k_{d1} & 0 & 0 \\ 0 & 0 & 1 & 0 \\ 0 & 0 & -k_{p2} & -k_{d2} \end{pmatrix}, \qquad (3.42)$$

where k_{p1}, k_{d1}, k_{p2}, and k_{d2} are chosen such that the matrix $\tilde{\tilde{A}}$ is Hurwitz. This boils down to a simple pole placement for the second order subsystems corresponding to the diagonal submatrices of $\tilde{\tilde{A}}$.

3.5.2.3 Learning-Based Auto-Tuning of the Feedback Gains

The goal is to track the desired output trajectory, despite the model uncertainties and with an auto-tuning of the feedback gain $k_{pi}, k_{di},\ i = 1, 2$. This control goal is useful in industrial applications where robot manipulators are used, for example, to move objects between two desired positions. Indeed, often the exact dynamics are not known and the manual tuning of the gains is time consuming. In this setting, an auto-tuning of the gains, with guaranteed stability of the system during the tuning, is an important result which is often referred to in the industry as auto-commissioning.

To do so, we first have to define a useful learning cost function which models the desired performance of the controller. In this particular case, because the final control goal is output trajectory tracking, we choose the following tracking cost function

$$Q(\gamma(\beta)) = \int_{(I-1)t_f}^{It_f} ey^T(t)C_1 ey(t)dt + \int_{(I-1)t_f}^{It_f} ey^{(1)^T}(t)C_2 ey^{(1)}(t)dt, \quad (3.43)$$

where $ey = (z_1, z_3)^T$, $C_1 = \text{diag}(c_{11}, c_{12}) > 0$, $C_2 = \text{diag}(c_{21}, c_{22}) > 0$, t_f is the finite time interval length of the desired trajectory, and $I = 1, 2, \ldots$ denotes the iteration index. β is the optimization variables vector, defined as $\beta = (\delta k_{p1}, \ \delta k_{d1}, \ \delta k_{p2}, \ \delta k_{d2}, \ \delta k)^T$, such as the feedback gains write as

$$k_{p1} = k_{p1_{\text{nominal}}} + \delta k_{p1},$$
$$k_{d1} = k_{d1_{\text{nominal}}} + \delta k_{d1},$$
$$k_{p2} = k_{p2_{\text{nominal}}} + \delta k_{p2}, \quad (3.44)$$
$$k_{d2} = k_{d2_{\text{nominal}}} + \delta k_{d2},$$
$$k = k_{\text{nominal}} + \delta k,$$

where $k_{pi_{\text{nominal}}}$, $k_{di_{\text{nominal}}}$, $i = 1, 2$ represent the nominal gains of the controller, that is, any starting gains that satisfy Assumption 3.6. Next, based on this learning cost function and the MES algorithms (3.18), (3.19), and (3.21), the variations of the gains are given by

$$\dot{x}_{k_{p1}} = a_{k_{p1}} \sin\left(\omega_1 t - \frac{\pi}{2}\right) Q(\gamma(\beta)),$$

$$\delta\hat{k}_{p1}(t) = x_{k_{p1}}(t) + a_{k_{p1}} \sin\left(\omega_1 t + \frac{\pi}{2}\right),$$

$$\dot{x}_{k_{d1}} = a_{k_{d1}} \sin\left(\omega_2 t - \frac{\pi}{2}\right) Q(\gamma(\beta)),$$

$$\delta\hat{k}_{d1}(t) = x_{k_{d1}}(t) + a_{k_{d1}} \sin\left(\omega_2 t + \frac{\pi}{2}\right),$$

$$\dot{x}_{k_{p2}} = a_{k_{p2}} \sin\left(\omega_3 t - \frac{\pi}{2}\right) Q(\gamma(\beta)),$$

$$\delta\hat{k}_{p2}(t) = x_{k_{p2}}(t) + a_{k_{p2}} \sin\left(\omega_3 t + \frac{\pi}{2}\right),$$

$$\dot{x}_{k_{d2}} = a_{k_{d2}} \sin\left(\omega_4 t - \frac{\pi}{2}\right) Q(\gamma(\beta)),$$

$$\delta\hat{k}_{d2}(t) = x_{k_{d2}}(t) + a_{k_{d2}} \sin\left(\omega_4 t + \frac{\pi}{2}\right),$$

$$\dot{x}_k = a_k \sin\left(\omega_5 t - \frac{\pi}{2}\right) Q(\gamma(\beta)),$$

$$\delta\hat{k}(t) = x_k(t) + a_k \sin\left(\omega_5 t + \frac{\pi}{2}\right),$$

$$\delta k_{pj}(t) = \delta\hat{k}_{pj}((I-1)t_f),$$

$$\delta k_{dj}(t) = \delta\hat{k}_{dj}((I-1)t_f), \ (I-1)t_f \le t < It_f, \ j \in \{1,2\}, \ I = 1,2,3,\dots$$

$$\delta k(t) = \delta\hat{k}((I-1)t_f), \ (I-1)t_f \le t < It_f, \ I = 1,2,3,\dots$$

$$(3.45)$$

where $a_{k_{p1}}$, $a_{k_{p2}}$, $a_{k_{d1}}$, $a_{k_{d2}}$, and a_k are the positive search amplitudes, and $\omega_j + \omega_q \ne \omega_r$, $p, q, r \in \{1, 2, 3, 4, 5\}$, for $p \ne q \ne r$.

3.5.2.4 Simulations Results
We choose as outputs' references the fifth-order polynomials

$$q_{1\text{ref}}(t) = q_{2\text{ref}}(t) = \sum_{i=0}^{5} a_i (t/t_f)^i,$$

where the a_i's have been computed to satisfy the boundary constraints

$$q_{i\text{ref}}(0) = 0, q_{i\text{ref}}(t_f) = q_f, \dot{q}_{i\text{ref}}(0) = \dot{q}_{i\text{ref}}(t_f) = 0, \ddot{q}_{i\text{ref}}(0) = \ddot{q}_{i\text{ref}}(t_f) = 0,$$

$i = 1, 2$, with $t_f = 1$ s, $q_f = 1.5$ rad.

We use the model nominal parameters summarized in Table 3.2. We assume bounded parametric uncertainties $\delta\eta_1$, $\delta\eta_2$, s.t., $\delta\eta_{\max} = 1$, and bounded additive disturbances $d(t)$, s.t., $d_{\max} = 10$. We apply the controller of Theorem 3.2, with $C_1 = \text{diag}(500, 100)$, $C_2 = \text{diag}(1000, 100)$, and

Table 3.2 Numerical values of the model parameters for the manipulator example

Parameter	Value
I_2	$\frac{5.5}{12}$ kg m^2
m_1	10.5 kg
m_2	5.5 kg
ℓ_1	1.1 m
ℓ_2	1.1 m
ℓ_{c_1}	0.5 m
ℓ_{c_2}	0.5 m
I_1	$\frac{11}{12}$ kg m^2
g	9.8 m/s^2

the learning frequencies are $\omega_1 = 7.5$ rad/s, $\omega_2 = 5.3$ rad/s, $\omega_3 = 10$ rad/s, $\omega_4 = 5.1$ rad/s, and $\omega_5 = 20$ rad/s.

In this test, we use constant search amplitudes $a_{k_{p1}} = 5.10^{-3}$, $a_{k_{d1}} = 10^{-2}/4$, $a_{k_{p2}} = 5.10^{-3}$, $a_{k_{d2}} = 10^{-2}/4$, and $a_k = 10^{-2}$. We start the gains' learning from the nominal values $k_{p1_{\text{nominal}}} = 10$, $k_{d1_{\text{nominal}}} = 5$, $k_{p2_{\text{nominal}}} = 10$, $k_{d2_{\text{nominal}}} = 5$, and $k_{\text{nominal}} = 0.1$. The controller (3.41), (3.43), (3.44), and (3.45) has been applied to the uncertain model of the robot. The obtained angular positions and velocities are reported in Fig. 3.11. We see clearly that without optimal tuning of the gains the system is stable (due to the robustness of the nonlinear controller), but the tracking performance is not good with the nominal gains. However, with the learning the performance is improved, as seen on the learning cost function in Fig. 3.12, which decreases over the iterations and reaches a very low values after 8 iterations only. The corresponding gains' learning iterations are reported in Fig. 3.13, which shows a fast convergence of the gains to an optimal feedback gain vector. Because, in this particular example, the test with constant search amplitudes led to very fast tuning and overall performance improvement, we did not try the time-varying search amplitudes algorithms, as we did in the electromagnetic actuator example.

3.6 CONCLUSION AND DISCUSSION OF OPEN PROBLEMS

In this chapter we have studied the problem of iterative feedback gains tuning for input–output linearization control with static-state feedback. First, we have used input–output linearization with static-state feedback method and "robustified" it with respect to bounded additive model uncertainties, using Lyapunov reconstruction techniques, to ensure uniform

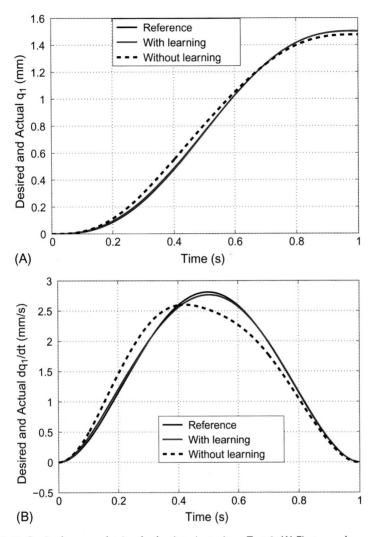

Fig. 3.11 Desired versus obtained robot's trajectories—Test 1. (A) First angular position (rad). (B) First angular velocity (rad/s).

(Continued)

boundedness of a tracking error vector. Second, we have complemented the input-output linearization controller with a model-free learning algorithm to iteratively auto-tune the control feedback gains and optimize a desired performance of the system. The learning algorithm used here is based on MES theory. The full controller, that is, the learning algorithm together with the robust controller forms an iterative input-output linearization-based controller, which auto-tune its gains. We have reported some

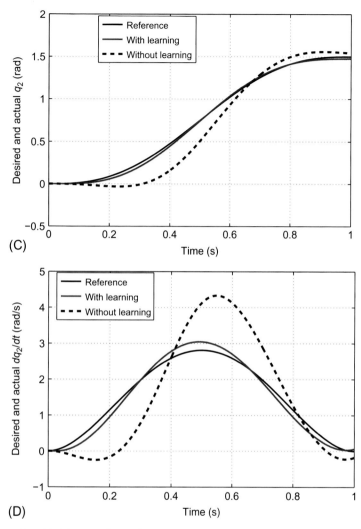

Fig. 3.11, Cont'd (C) Second angular position (rad). (D) Second angular velocity (rad/s).

numerical results obtained on an electromagnetic actuator example, as well as a two-link manipulator example. Open problems concern improving the convergence rate by using different MES algorithms with semiglobal convergence properties, see Tan et al. (2006), Noase et al. (2011), and Khong et al. (2013), and extending this work to different types of model-free learning algorithms, for example, machine learning algorithms.

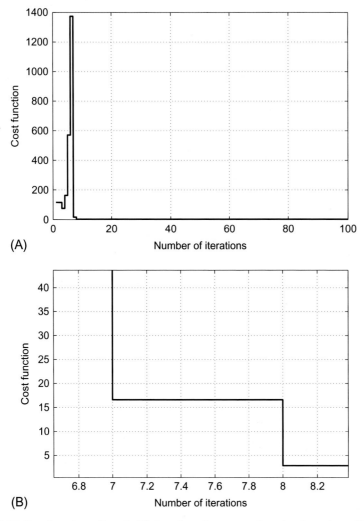

Fig. 3.12 Cost function—Test 1. (A) Cost function versus learning iterations. (B) Cost function versus learning iterations—Zoom.

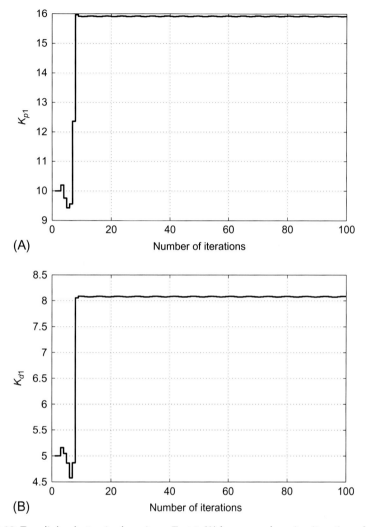

Fig. 3.13 Two-link robot gains learning—Test 1. (A) k_{p1} versus learning iterations. (B) k_{d1} versus learning iterations.

(Continued)

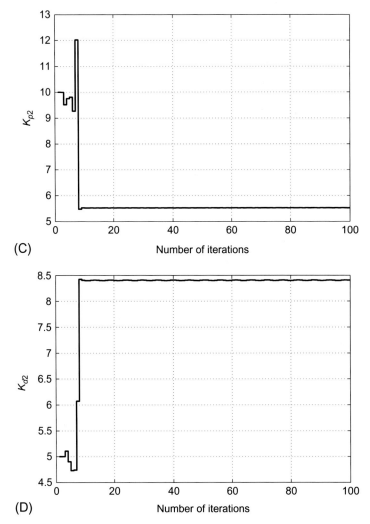

Fig. 3.13, Cont'd (C) k_{p2} versus learning iterations. (D) k_{d2} versus learning iterations.

(Continued)

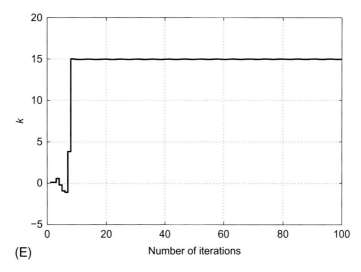

Fig. 3.13, Cont'd (E) *k* versus learning iterations.

REFERENCES

Ariyur, K.B., Krstic, M., 2002. Multivariable extremum seeking feedback: analysis and design. In: Proceedings of the Mathematical Theory of Networks and Systems, South Bend, IN.

Benosman, M., 2014. Multi-parametric extremum seeking-based auto-tuning for robust input-output linearization control. In: IEEE, Conference on Decision and Control, Los Angeles, CA, pp. 2685–2690.

Benosman, M., Atinc, G., 2013. Multi-parametric extremum seeking-based learning control for electromagnetic actuators. In: IEEE, American Control Conference, Washington, DC, pp. 1914–1919.

Benosman, M., Lum, K.Y., 2010. Passive actuators' fault tolerant control for affine nonlinear systems. IEEE Trans. Control Syst. Technol. 18 (1), 152–163.

DeBruyne, F., Anderson, B., Gevers, M., Linard, N., 1997. Iterative controller optimization for nonlinear systems. In: IEEE, Conference on Decision and Control, pp. 3749–3754.

Haddad, W.M., Chellaboind, V., Nersesov, S.G., 2006. Impulsive and Hybrid Dynamical Systems: Stability, Dissipativity, and Control. Princeton University Press, Princeton.

Hale, J., 1977. Theory of Functional Differential Equations, Applied Mathematical Sciences, vol. 3. Springer-Verlag, New York.

Hjalmarsson, H., 1998. Control of nonlinear systems using iterative feedback tuning. In: IEEE, Conference on Decision and Control, pp. 2083–2087.

Hjalmarsson, H., 2002. Iterative feedback tuning—an overview. Int. J. Adapt. Control Signal Process. 16 (5), 373–395. doi:10.1002/acs.714.

Hjalmarsson, H., Gunnarsson, S., Gevers, M., 1994. A convergent iterative restricted complexity control design scheme. In: IEEE, Conference on Decision and Control, pp. 1735–1740.

Hjalmarsson, H., Gevers, M., Gunnarsson, S., Lequin, O., 1998. Iterative feedback tuning: theory and applications. IEEE Control Syst. 18 (4), 26–41.

Isidori, A., 1989. Nonlinear Control Systems, second ed., Communications and Control Engineering Series. Springer-Verlag, Berlin.

Khalil, H., 1996. Nonlinear Systems, second ed. Macmillan, New York.

Khong, S.Z., Nesic, D., Tan, Y., Manzie, C., 2013. Unified frameworks for sampled-data extremum seeking control: global optimization and multi-unit systems. Automatica 49, 2720–2733.

Killingsworth, N., Kristic, M., 2006. PID tuning using extremum seeking. In: IEEE Control Systems Magazine, pp. 1429–1439.

Koszalka, L., Rudek, R., Pozniak-Koszalka, I., 2006. An idea of using reinforcement learning in adaptive control systems. In: International Conference on Networking, International Conference on Systems and International Conference on Mobile Communications and Learning Technologies, 2006, ICN/ICONS/MCL 2006, pp. 190–196.

Lequin, O., Gevers, M., Mossberg, M., Bosmans, E., Triest, L., 2003. Iterative feedback tuning of PID parameters: comparison with classical tuning rules. Control Eng. Pract. 11 (9), 1023–1033.

Moase, W., Manzie, C., Brear, M., 2009. Newton-like extremum seeking part I: theory. In: IEEE, Conference on Decision and Control, pp. 3839–3844.

Noase, W., Tan, Y., Nesic, D., Manzie, C., 2011. Non-local stability of a multi-variable extremum-seeking scheme. In: IEEE, Australian Control Conference, pp. 38–43.

Peterson, K., Stefanopoulou, A., 2004. Extremum seeking control for soft landing of electromechanical valve actuator. Automatica 40, 1063–1069.

Rotea, M.A., 2000. Analysis of multivariable extremum seeking algorithms. In: IEEE, American Control Conference, pp. 433–437.

Scheinker, A., 2013. Simultaneous stabilization of and optimization of unknown time-varying systems. In: IEEE, American Control Conference, pp. 2643–2648.

Sjoberg, J., Agarwal, M., 1996. Model-free repetitive control design for nonlinear systems. In: IEEE, Conference on Decision and Control, pp. 2824–2829.

Sjoberg, J., Bruyne, F.D., Agarwal, M., Anderson, B., Gevers, M., Kraus, F., Linard, N., 2003. Iterative controller optimization for nonlinear systems. Control Eng. Pract. 11, 1079–1086.

Spong, M.W., 1992. On the robust control of robot manipulators. IEEE Trans. Autom. Control 37 (11), 1782–2786.

Tan, Y., Nesic, D., Mareels, I., 2006. On non-local stability properties of extremum seeking control. Automatica 42, 889–903.

Tan, Y., Nesic, D., Mareels, I., 2008. On the dither choice in extremum seeking control. Automatica 44, 1446–1450.

Wang, Y., Stefanopoulou, A., Haghgooie, M., Kolmanovsky, I., Hammoud, M., 2000. Modelling of an electromechanical valve actuator for a camless engine. In: Fifth International Symposium on Advanced Vehicle Control, Number 93.

Wang, Y., Gao, F., Doyle III, F.J., 2009. Survey on iterative learning control, repetitive control, and run-to-run control. J. Process Control 19, 1589–1600.

Ziegler, J., Nichols, N., 1942. Optimum settings for automatic controllers. Trans. ASME 64, 759–768.

CHAPTER 4

Extremum Seeking-Based Indirect Adaptive Control

4.1 INTRODUCTION

Extremum seeking (ES) is a well-known approach by which one can search for the extremum of a cost function associated with a given process performance without the need of a precise modeling of the process, see Ariyur and Krstic (2002, 2003) and Nesic (2009). Several ES algorithms have been proposed in: Krstic (2000), Rotea (2000), Ariyur and Krstic (2002, 2003), Tan et al. (2006), Nesic (2009), Guay et al. (2013), and Scheinker (2013), and many applications of ES algorithms have been reported, see Zhang et al. (2003), Hudon et al. (2008), Zhang and Ordóñez (2012), and Benosman and Atinc (2013a,c). Many papers have been dedicated to analyzing the ES algorithms, convergence when applied to static or dynamic known maps, see Krstic and Wang (2000), Rotea (2000), Teel and Popovic (2001), and Ariyur and Krstic (2003); however, much fewer papers deals with the use of ES in the case of static or dynamic uncertain maps. The case of ES applied to an uncertain static and dynamic mapping was investigated in Nesic et al. (2013), where the authors considered systems with constant parameter uncertainties. However, in Nesic et al. (2013), the authors used ES to optimize a given performance (via optimizing a performance function), and complemented the ES algorithm with classical model-based filters/estimators to estimate the states and the unknown constant parameters of the system. This is different from the approach that we are presenting here, where the ES is used to improve a given control performance, and at the same time estimate the model uncertainties.

As seen in Chapter 2, historically the control of uncertain maps is dealt with using classical adaptive control. Indeed, classical model-based adaptive control deals with controlling partially unknown processes based on their uncertain model, that is, controlling plants with parameter uncertainties. One can classify classical adaptive methods into two main approaches: direct

Learning-Based Adaptive Control
http://dx.doi.org/10.1016/B978-0-12-803136-0.00004-X

approaches, where the controller is updated to adapt to the process; and indirect approaches, where the model is updated to better reflect the actual process. Many adaptive methods have been proposed over the years for linear and nonlinear systems. We could not possibly cite here all the design and analysis results that have been reported; instead, we refer the reader to, for example, Krstic et al. (1995) and Landau et al. (2011) and the references therein for more details. What we want to underline here is that these results in classical adaptive control are mainly based on the structure of the model of the system, for example, linear versus nonlinear, with linear uncertainties parameterization versus nonlinear parameterization, etc.

Another adaptive control paradigm is the one which uses learning schemes to estimate the uncertain part of the process. Indeed, in this paradigm the learning-based controller, based on machine learning theory, neural networks, fuzzy systems, and so on, is trying either to estimate the parameters of an uncertain model, or the structure of a deterministic or stochastic function representing part or totality of the model. Several results have been proposed in this area as well; we refer the reader to, for example, Wang and Hill (2006) and the references therein for more details.

We want to concentrate in this chapter on the use of ES theory in the learning-based adaptive control paradigm. Indeed, several results have been developed in this direction, see Guay and Zhang (2003), Zhang et al. (2003), Adetola and Guay (2007), Ariyur et al. (2009), Hudon et al. (2008), Haghi and Ariyur (2011), and Benosman and Atinc (2013a,c). For instance, in Guay and Zhang (2003) and Adetola and Guay (2007), an ES-based controller was proposed for nonlinear affine systems with linear parameter uncertainties. The controller drives the states of the system to unknown optimal states that optimize a desired objective function. The ES controller used in Guay and Zhang (2003) and Adetola and Guay (2007) is not model-free in the sense that it is based on the known part of the model; that is, it is designed based on the objective function and the nonlinear model structure. A similar approach is used in Zhang et al. (2003) and Hudon et al. (2008) when dealing with more specific examples. In Haghi and Ariyur (2011) and Ariyur et al. (2009) the authors used a model-free ES, which is based on a desired cost function, to estimate parameters of a linear state feedback to compensate for unknown parameters for linear systems. In Atinc and Benosman (2013) the problem of adaptive robust control of electromagnetic actuators was studied, where ES was used to estimate the unknown model parameters. An extension to the general case of nonlinear systems was proposed in Benosman (2014a,b). Finally, in Benosman and

Xia (2015) the strong assumption about the existence of an input–to–state stable (ISS) feedback controller, used in Benosman (2014a,b), was relaxed and a constructive proof to design such an ISS feedback for the particular case of nonlinear systems affine in the control was proposed.

In this context we present in this chapter some ES-based indirect adaptive controllers for nonlinear systems. The results reported here are based on the work of the author, introduced in some of the references quoted previously. The idea is based on a modular design. We first design a feedback controller which makes the closed-loop tracking error dynamic ISS (or a similar small gain property) with respect to the estimation errors, and then complement this ISS controller with a model-free ES algorithm that can minimize a desired cost function by estimating the unknown parameters of the model. The modular design simplifies the analysis of the total controller, that is, ISS controller plus ES estimation algorithm.

This chapter is organized as follows: In Section 4.2 we recall some notations and definitions that will be used in this chapter. In Section 4.3 we present the first indirect adaptive control approach, namely, the ES-based learning adaptive controller for constant parametric model uncertainties. In Section 4.4 we study the case of time-varying parametric model uncertainties using time-varying ES-based techniques. Section 4.5 is dedicated to the specific class of nonlinear models affine in the control. We study in Section 4.6 two mechatronic examples, namely, electromagnetic actuators and rigid robot manipulators. We apply the proposed techniques to these problems and report some numerical results. Finally, summarizing remarks and open problems are discussed in Section 4.7.

4.2 BASIC NOTATIONS AND DEFINITIONS

Throughout the chapter we will use $\| \cdot \|$ to denote the Euclidean norm; that is, for $x \in \mathbb{R}^n$ we have $\|x\| = \sqrt{x^T x}$. We will use $(\dot{\cdot})$ for the short notation of time derivative. We denote by $\mathcal{C}^k$ functions that are k times differentiable. A function is said to be analytic in a given set if it admits a convergent Taylor series approximation in some neighborhood of every point of the set. A continuous function $\alpha: [0, a) \rightarrow [0, \infty)$ is said to belong to class $\mathcal{K}$ if it is strictly increasing and $\alpha(0) = 0$. A continuous function $\beta: [0, a) \times [0, \infty) \rightarrow [0, \infty)$ is said to belong to class $\mathcal{KL}$ if, for each fixed s, the mapping $\beta(r, s)$ belongs to class $\mathcal{K}$ with respect to r and, for each fixed r, the mapping $\beta(r, s)$ is decreasing with respect to s and $\beta(r, s) \rightarrow 0$

as $s \to \infty$. The Frobenius norm of a matrix $A \in \mathbb{R}^{m \times n}$, with elements a_{ij}, is defined as $\|A\|_F \triangleq \sqrt{\sum_{i=1}^{n} \sum_{j=1}^{n} |a_{ij}|^2}$. We use the following norm properties (cf. Chapter 1):

-   for any $x \in \mathbb{R}^n, A \in \mathbb{R}^{m \times n}, \|Ax\| \leq \|A\|_F \|x\|,$
-   for any $x, y \in \mathbb{R}^n, \|x\| - \|y\| \leq \|x - y\|,$
-   for any $x, y \in \mathbb{R}^n, x^T y \leq \|x\| \|y\|.$

Given $x \in \mathbb{R}^m$, the signum function for a vector x is defined as

$$\text{sign}(x) \triangleq [\text{sign}(x_1), \ \text{sign}(x_2), \ldots, \text{sign}(x_m)]^T,$$

where x_i denotes the ith $(1 \leq i \leq m)$ element of x. Thus we have $x^T \text{sign}(x) = \|x\|_1$.

For an $n \times n$ matrix P, we write $P > 0$ if it is positive definite. Similarly, we write $P < 0$ if it is negative definite. We use $\text{diag}\{A_1, A_2, \ldots, A_n\}$ to denote a diagonal block matrix with n blocks. For a matrix B, we denote $B(i, j)$ as the (i, j) element of B. We denote by I_n the identity matrix or simply I if the dimension is clear from the context; we denote $\mathbf{0}$ the vector $[0, \ldots, 0]^T$ of appropriate dimension.

Let us now introduce a few more definitions which will be used in the remainder of this chapter. First, we introduce a fundamental ISS result.

Consider the system

$$\dot{x} = f(t, x, u), \tag{4.1}$$

where $f: [0, \infty) \times \mathbb{R}^n \times \mathbb{R}^{na} \to \mathbb{R}^n$ is piecewise continuous in t and locally Lipschitz in x and u, uniformly in t. The input $u(t)$ is a piecewise continuous, bounded function of t for all $t \geq 0$.

Definition 4.1 (Khalil, 2002). *The system (4.1) is said to be ISS if there exist a class $\mathcal{KL}$ function β and a class $\mathcal{K}$ function γ such that for any initial state $x(t_0)$ and any bounded input $u(t)$, the solution $x(t)$ exists for all $t \geq t_0$ and satisfies*

$$\|x(t)\| \leq \beta(\|x(t_0)\|, t - t_0) + \gamma \left(\sup_{t_0 \leq \tau \leq t} \|u(\tau)\| \right).$$

Theorem 4.1 (Khalil, 2002). *Let $V: [0, \infty) \times \mathbb{R}^n \to \mathbb{R}$ be a continuously differentiable function such that*

$$\alpha_1(\|x\|) \leq V(t, x) \leq \alpha_2(\|x\|),$$

$$\frac{\partial V}{\partial t} + \frac{\partial V}{\partial x} f(t, x, u) \leq -W(x), \quad \forall \|x\| \geq \rho(\|u\|) > 0 \tag{4.2}$$

for all $(t, x, u) \in [0, \infty) \times \mathbb{R}^n \times \mathbb{R}^{na}$, where α_1, α_2 are class $\mathcal{K}_\infty$ functions, ρ is a class $\mathcal{K}$ function, and $W(x)$ is a continuous positive definite function on $\mathbb{R}^n$. Then, the system (4.1) is ISS.

Next, we recall the definition of a weaker property than ISS, referred to as local integral input-to-state stability.

Definition 4.2 (Ito and Jiang, 2009). *Consider the system*

$$\dot{x} = f(t, x, u), \tag{4.3}$$

where $x \in \mathcal{D} \subseteq \mathbb{R}^n$ such that $0 \in \mathcal{D}$, and $f \colon [0, \infty) \times \mathcal{D} \times \mathcal{D}_u \to \mathbb{R}^n$ is piecewise continuous in t and locally Lipschitz in x and u, uniformly in t. The inputs are assumed to be measurable and locally bounded functions $u \colon \mathbb{R}_{\geq 0} \to \mathcal{D}_u \subseteq \mathbb{R}^{na}$. Given any control $u \in \mathcal{D}_u$ and any $\xi \in \mathcal{D}_0 \subseteq \mathcal{D}$, there is a unique maximal solution of the initial value problem $\dot{x} = f(t, x, u)$, $x(t_0) = \xi$. Without loss of generality, assume $t_0 = 0$. The unique solution is defined on some maximal open interval, and it is denoted by $x(\cdot, \xi, u)$. System (4.3) is locally integral input-to-state stable (LiISS) if there exist functions $\alpha, \gamma \in \mathcal{K}$ and $\beta \in \mathcal{KL}$ such that, for all $\xi \in \mathcal{D}_0$ and all $u \in \mathcal{D}_u$, the solution $x(t, \xi, u)$ is defined for all $t \geq 0$ and

$$\alpha(\|x(t, \xi, u)\|) \leq \beta(\|\xi\|, t) + \int_0^t \gamma(\|u(s)\|) ds, \tag{4.4}$$

for all $t \geq 0$. Equivalently, system (4.3) is LiISS if and only if there exist functions $\beta \in \mathcal{KL}$ and $\gamma_1, \gamma_2 \in \mathcal{K}$ such that

$$\|x(t, \xi, u)\| \leq \beta(\|\xi\|, t) + \gamma_1 \left(\int_0^t \gamma_2(\|u(s)\|) ds \right), \tag{4.5}$$

for all $t \geq 0$, all $\xi \in \mathcal{D}_0$, and all $u \in \mathcal{D}_u$.

Remark 4.1. *We want to underline here that the two definitions recalled previously about ISS and LiISS are only two examples of properties which can be used in the context of the adaptive modular design presented here. Indeed, the main idea is to ensure a small gain property between a given input and given states. The small gain property can be formulated in term of ISS, LiISS, or other definitions, for example, ISS with decay rate p (ISS(p)) in Malisoff and Mazenc (2005), and so on.*

Another property that will be used in this chapter is ϵ-semiglobal practical uniform ultimate boundedness with ultimate bound $\delta((\epsilon - \delta)$-SPUUB), which is defined next.

Definition 4.3 $(((\epsilon - \delta)$-SPUUB); Scheinker, 2013). *Consider the system*

$$\dot{x} = f^\epsilon(t, x), \tag{4.6}$$

with $\phi^\epsilon(t, t_0, x_0)$ being the solution of Eq. (4.6) starting from the initial condition $x(t_0) = x_0$. Then, the origin of Eq. (4.6) is said to be (ϵ, δ)-SPUUB if it satisfies the following three conditions:

1. *(ϵ, δ)-Uniform stability: For every $c_2 \in]\delta, \infty[$, there exist $c_1 \in]0, \infty[$ and $\hat{\epsilon} \in]0, \infty[$ such that for all $t_0 \in \mathbb{R}$ and for all $x_0 \in \mathbb{R}^n$ with $\|x_0\| < c_1$ and for all $\epsilon \in]0, \hat{\epsilon}[$,*

$$\|\phi^\epsilon(t, t_0, x_0)\| < c_2, \quad \forall t \in [t_0, \infty[.$$

2. *(ϵ, δ)-Uniform ultimate boundedness: For every $c_1 \in]0, \infty[$ there exist $c_2 \in]\delta, \infty[$ and $\hat{\epsilon} \in]0, \infty[$ such that for all $t_0 \in \mathbb{R}$ and for all $x_0 \in \mathbb{R}^n$ with $\|x_0\| < c_1$ and for all $\epsilon \in]0, \hat{\epsilon}[$,*

$$\|\phi^\epsilon(t, t_0, x_0)\| < c_2, \quad \forall t \in [t_0, \infty[.$$

3. *(ϵ, δ)-Global uniform attractivity: For all $c_1, c_2 \in (\delta, \infty)$ there exist $T \in]0, \infty[$ and $\hat{\epsilon} \in]0, \infty[$ such that for all $t_0 \in \mathbb{R}$ and for all $x_0 \in \mathbb{R}^n$ with $\|x_0\| < c_1$ and for all $\epsilon \in]0, \hat{\epsilon}[$,*

$$\|\phi^\epsilon(t, t_0, x_0)\| < c_2, \quad \forall t \in [t_0 + T, \infty[.$$

We now have all the needed definitions to present the first part of this chapter, regarding the modular ES-based adaptive control for nonlinear systems. This first part is presented for general nonlinear models under somehow strong assumptions; however, later on we will weaken some of the assumptions when dealing with a more specific class of nonlinear systems.

4.3 ES-BASED INDIRECT ADAPTIVE CONTROLLER FOR THE CASE OF GENERAL NONLINEAR MODELS WITH CONSTANT MODEL UNCERTAINTIES

Consider the model (4.3) with an additional variable representing constant parametric uncertainties $\Delta \in \mathbb{R}^p$

$$\dot{x} = f(t, x, \Delta, u). \tag{4.7}$$

We associate with Eq. (4.7) the output vector

$$y = h(x), \tag{4.8}$$

where $h: \mathbb{R}^n \to \mathbb{R}^m$.

The control objective here is for y to asymptotically track a desired smooth vector time-dependent trajectory $y_{\text{ref}}: [0, \infty) \to \mathbb{R}^h$.

Let us now define the output tracking error vector as

$$e_y(t) = y(t) - y_{\text{ref}}(t). \tag{4.9}$$

We then state the following assumptions.

Assumption 4.1. *There exists a robust control feedback $u_{iss}(t, x, \hat{\Delta}): \mathbb{R} \times \mathbb{R}^n \times \mathbb{R}^p \to \mathbb{R}^{n_a}$, with $\hat{\Delta}$ being the dynamic estimate of the uncertain vector Δ, such that the closed-loop error dynamics*

$$\dot{e}_y = f_{e_y}(t, e_y, e_\Delta), \tag{4.10}$$

is LiISS from the input vector $e_\Delta = \Delta - \hat{\Delta}$ to the state vector e_y.

Remark 4.2. *Assumption 4.1 might seem too general; however, several control approaches can be used to design a controller u_{iss} rendering an uncertain system LiISS. For instance, the backstepping control approach has been shown to achieve such a property for parametric strict-feedback systems, see for example, Krstic et al. (1995).*

Let us define now the following cost function

$$Q(\hat{\Delta}) = F(e_y(\hat{\Delta})), \tag{4.11}$$

where $F: \mathbb{R}^h \to \mathbb{R}$, $F(0) = 0$, $F(e_y) > 0$ for $e_y \neq 0$. We need the following additional assumptions on Q.

Assumption 4.2. *The cost function Q has a local minimum at $\hat{\Delta}^* = \Delta$.*

Assumption 4.3. *The initial error $e_\Delta(t_0)$ is sufficiently small; that is, the original parameter estimate vector $\hat{\Delta}$ is close enough to the actual parameter vector Δ.*

Assumption 4.4. *The cost function is analytic and its variation with respect to the uncertain variables is bounded in the neighborhood of Δ^*, that is, $\|\frac{\partial Q}{\partial \Delta}(\tilde{\Delta})\| \leq \xi_2$, $\xi_2 > 0$, $\tilde{\Delta} \in \mathcal{V}(\Delta^*)$, where $\mathcal{V}(\Delta^*)$ denotes a compact neighborhood of Δ^*.*

Remark 4.3. *Assumption 4.2 simply means that we can consider that Q has at least a local minimum at the true values of the uncertain parameters.*

Remark 4.4. *Assumption 4.3 indicates that our result will be of local nature, meaning that our analysis holds in a small neighborhood of the actual values of the parameters.*

We can now present the following theorem.

Theorem 4.2. *Consider the system (4.7), (4.8), with the cost function (4.11), and the controller u_{iss}, where $\hat{\Delta}$ is estimated with the multiparametric extremum-seeking (MES) algorithm*

$$\dot{z}_i = a_i \sin\left(\omega_i t + \frac{\pi}{2}\right) Q(\hat{\Delta}),$$
$$\hat{\Delta}_i = z_i + a_i \sin\left(\omega_i t - \frac{\pi}{2}\right), \quad i \in \{1, \ldots, p\} \tag{4.12}$$

with $\omega_i \neq \omega_j$, $\omega_i + \omega_j \neq \omega_k$, $i, j, k \in \{1, \ldots, p\}$, and $\omega_i > \omega^*, \forall i \in \{1, \ldots, p\}$, with ω^* large enough. Then, under Assumptions 4.1–4.4, the norm of the error vector e_y admits the following bound

$$\|e_y(t)\| \leq \beta(\|e_y(0)\|, t) + \alpha \left(\int_0^t \gamma(\tilde{\beta}(\|e_\Delta(0)\|, s) + \|e_\Delta\|_{\max}) ds \right),$$

where $\|e_\Delta\|_{\max} = \frac{\xi_1}{\omega_0} + \sqrt{\sum_{i=1}^{i=p} a_i^2}$, $\xi_1, \xi_2 > 0$, $\omega_0 = \max_{i \in \{1, \ldots, p\}} \omega_i$, $\alpha \in \mathcal{K}$, $\beta \in \mathcal{KL}, \tilde{\beta} \in \mathcal{KL}$, and $\gamma \in \mathcal{K}$.

Proof. Consider the system (4.7), (4.8), then under Assumption 4.1, the controller u_{iss} ensures that the tracking error dynamic (4.10) is LiISS between the input e_Δ and the state vector e_y, which by Definition 4.1 implies that there exist functions $\alpha \in \mathcal{K}$, $\beta \in \mathcal{KL}$, and $\gamma \in \mathcal{K}$, such that for all $e_y(0) \in \mathcal{D}_{e_y}$ and $e_\Delta \in \mathcal{D}_{e_\Delta}$ the norm of the error vector e_Δ admits the following bound

$$\|e_y(t)\| \leq \beta(\|e_y(0)\|, t) + \alpha \left(\int_0^t \gamma(\|e_\Delta\|) ds \right) \qquad (4.13)$$

for all $t \geq 0$.

Next we need to evaluate the bound on the estimation vector $\tilde{\Delta}$; to do so we use the results presented in Rotea (2000). First, based on Assumption 4.5, the cost function is locally Lipschitz, that is, $\exists \eta_1 > 0$, s.t., $|Q(\Delta_1) - Q(\Delta_2)| \leq \eta_1 \|\Delta_1 - \Delta_2\|$, $\forall \Delta_1, \Delta_2 \in \mathcal{V}(\Delta^*)$. Furthermore, because Q is analytic it can be approximated locally in $\mathcal{V}(\Delta^*)$ with a quadratic function, for example, Taylor series up to second order. Based on this and on Assumptions 4.2 and 4.3, we can write the following bound (Rotea, 2000, pp. 436–437):

$$\|e_\Delta(t)\| - \|d(t)\| \leq \|e_\Delta(t) - d(t)\| \leq \tilde{\beta}(\|e_\Delta(0)\|, t) + \frac{\xi_1}{\omega_0},$$

$$\Rightarrow \|e_\Delta(t)\| \leq \tilde{\beta}(\|e_\Delta(0)\|, t) + \frac{\xi_1}{\omega_0} + \|d(t)\|,$$

$$\Rightarrow \|e_\Delta(t)\| \leq \tilde{\beta}(\|e_\Delta(0)\|, t) + \frac{\xi_1}{\omega_0} + \sqrt{\sum_{i=1}^{i=p} a_i^2},$$

with $\tilde{\beta} \in \mathcal{KL}$, $\xi_1 > 0$, $t \geq 0$, $\omega_0 = \max_{i \in \{1, \ldots, p\}} \omega_i$, and $d(t) = [a_1 \sin(\omega_1 t + \frac{\pi}{2}), \ldots, a_p \sin(\omega_p t + \frac{\pi}{2})]^T$, which together with the bound (4.13) completes the proof.

So far we have considered the case where the model uncertainties are constants. This case can appear in practical applications when a physical part of the system undergoes an abrupt change, for example, a new passenger stepping in an elevator car, which changes the mass of the system almost instantly. In other situations, however, changes in the system's parameters might happen over time, leading to time-varying uncertainties. One well-known example of such changes is the phenomenon of system aging. Indeed, in case of system aging, some parts of the system age from the repetitive use of the system over time, and some of their physical "constants" start drifting over time. These parameters can be modeled as time-varying uncertainties. Furthermore, due to the inherent slow dynamics of aging (in general), the time variation of the uncertainties is slow, which makes this case very suitable for adaptive control techniques. We focus in the next section on the study of nonlinear models with time-varying parametric uncertainties.

4.4 ES-BASED INDIRECT ADAPTIVE CONTROLLER FOR GENERAL NONLINEAR MODELS WITH TIME-VARYING MODEL UNCERTAINTIES

Consider the system (4.7) with time-varying parametric uncertainties $\Delta(t)\colon \mathbb{R} \to \mathbb{R}^p$, to which we associate the output vector (4.8). We consider the same control objective here, which is for y to asymptotically track a desired smooth vector time-dependent trajectory $y_{\text{ref}}\colon [0, \infty) \to \mathbb{R}^m$.

Let us define now the following cost function

$$Q(\hat{\Delta}, t) = F(e_y(\hat{\Delta}), t) \tag{4.14}$$

where $F\colon \mathbb{R}^m \times \mathbb{R}^+ \to \mathbb{R}^+$, $F(0, t) = 0$, $F(e_y, t) > 0$ for $e_y \neq 0$.

In this case, we need the following assumptions on Q.

Assumption 4.5. $|\frac{\partial Q(\hat{\Delta}, t)}{\partial t}| < \rho_Q$, $\forall t \in \mathbb{R}^+, \forall \hat{\Delta} \in \mathbb{R}^p$.

We can now state the following result.

Theorem 4.3. *Consider the system (4.7), (4.8), with the cost function (4.14), and a controller u_{iss} satisfying Assumption 4.1, where $\hat{\Delta}$ is estimated with the MES algorithm*

$$\dot{\hat{\Delta}}_i = a\sqrt{\omega_i}\cos(\omega_i t) - k\sqrt{\omega_i}\sin(\omega_i t)Q(\hat{\Delta}), \quad i \in \{1, \dots, p\} \tag{4.15}$$

with $a > 0, k > 0$, $\omega_i \neq \omega_j, i, j, k \in \{1, \dots, p\}$, and $\omega_i > \omega^*, \forall i \in \{1, \dots, p\}$, with ω^* large enough. Then, under Assumptions 4.2, and 4.5, the norm of the error vector e_y admits the following bound

$$\|e_y(t)\| \leq \beta(\|e_y(0)\|, t) + \alpha \left(\int_0^t \gamma(\|e_\Delta(s)\|) ds \right),$$

where $\alpha \in \mathcal{K}$, $\beta \in \mathcal{KL}$, $\gamma \in \mathcal{K}$, and $\|e_\Delta\|$ satisfies:

1. $(\frac{1}{\omega}, d)$-uniform stability: For every $c_2 \in]d, \infty[$, there exist $c_1 \in]0, \infty[$ and $\hat{\omega} > 0$ such that for all $t_0 \in \mathbb{R}$ and for all $x_0 \in \mathbb{R}^n$ with $\|e_\Delta(0)\| < c_1$ and for all $\omega > \hat{\omega}$,

$$\|e_\Delta(t, e_\Delta(0))\| < c_2, \quad \forall t \in [t_0, \infty[.$$

2. $(\frac{1}{\omega}, d)$-uniform ultimate boundedness: For every $c_1 \in]0, \infty[$ there exist $c_2 \in]d, \infty[$ and $\hat{\omega} > 0$ such that for all $t_0 \in \mathbb{R}$ and for all $x_0 \in \mathbb{R}^n$ with $\|e_\Delta(0)\| < c_1$ and for all $\omega > \hat{\omega}$,

$$\|e_\Delta(t, e_\Delta(0))\| < c_2, \quad \forall t \in [t_0, \infty[.$$

3. $(\frac{1}{\omega}, d)$-global uniform attractivity: For all $c_1, c_2 \in (d, \infty)$ there exist $T \in]0, \infty[$ and $\hat{\omega} > 0$ such that for all $t_0 \in \mathbb{R}$ and for all $x_0 \in \mathbb{R}^n$ with $\|e_\Delta(0)\| < c_1$ and for all $\omega > \hat{\omega}$,

$$\|e_\Delta(t, e_\Delta(0))\| < c_2, \quad \forall t \in [t_0 + T, \infty[,$$

where d is given by $d = \min\{r \in]0, \infty[: \Gamma_H \subset B(\Delta, r)\}$, with $\Gamma_H = \left\{ \hat{\Delta} \in \mathbb{R}^n: \left\| \frac{\partial Q(\hat{\Delta}, t)}{\partial \hat{\Delta}} \right\| < \sqrt{\frac{2\rho_Q}{ka\beta_0}} \right\}, 0 < \beta_0 \leq 1$, and $B(\Delta, r) = \{\hat{\Delta} \in \mathbb{R}^n: \|\hat{\Delta} - \Delta\| < r\}$.

Remark 4.5. Theorem 4.3 shows that the estimation error is bounded by a constant c_2 which can be tightened by making the constant d small. The d constant can be tuned by tuning the cardinal of the set Γ_H, which in turn can be made small by choosing large values for the coefficients a and k of the MES algorithm (4.15).

Proof. Consider the system (4.7), (4.8), then under Assumption 4.1, the controller u_{iss} ensures that the tracking error dynamic (4.10) is LiISS between the input e_Δ and the state vector e_y, which, by Definition 4.1, implies that there exist functions $\alpha \in \mathcal{K}$, $\beta \in \mathcal{KL}$, and $\gamma \in \mathcal{K}$, such that, for all $e(0) \in \mathcal{D}_e$ and $e_\Delta \in \mathcal{D}_{e_\Delta}$, the norm of the error vector e_Δ admits the following bound

$$\|e_y(t)\| \leq \beta(\|e_y(0)\|, t) + \alpha \left(\int_0^t \gamma(\|e_\Delta\|) ds \right), \tag{4.16}$$

for all $t \geq 0$.

Now we need to evaluate the bound of the estimation vector $\tilde{\Delta}$; to do so we use the results presented in Scheinker (2013). Indeed, based on

Theorem 3 of Scheinker (2013), we can conclude under Assumption 4.5 that the estimator (4.15) makes the local optimum of Q, that is, $\Delta^* = \Delta$ (see Assumption 4.2), $(\frac{1}{\omega}, d)$-SPUUB, where $d = \min\{r \in]0, \infty[: \Gamma_H \subset B(\Delta, r)\}$, with $\Gamma_H = \left\{ \hat{\Delta} \in \mathbb{R}^n : \left| \frac{\partial Q(\hat{\Delta}, t)}{\partial \hat{\Delta}} \right| < \sqrt{\frac{2\rho_Q}{ka\beta_0}} \right\}$, $0 < \beta_0 \leq 1$, and $B(\Delta, r) = \{ \hat{\Delta} \in \mathbb{R}^n : \| \hat{\Delta} - \Delta \| < r \}$, which by Definition 4.2 implies that $\|e_\Delta\|$ satisfies the three conditions: $(\frac{1}{\omega}, d)$-uniform stability, $(\frac{1}{\omega}, d)$-uniform ultimate boundedness, and $(\frac{1}{\omega}, d)$-global uniform attractivity.

   Remark 4.6. *The upper bounds of the estimated parameters used in Theorems 4.2 and 4.3 are correlated to the choice of the extremum-seeking algorithm (4.12), (4.15); however, these bounds can be easily changed by using other ES algorithms, see Noase et al. (2011). This is due to the modular design of the controller, which uses the LiISS robust part to ensure boundedness of the error dynamics and the learning part to improve the tracking performance.*

   The models considered up to this point are in a very general nonlinear form. To be able to obtain more constructive results we need to consider a more specific class of systems. One specific, yet general enough, class of systems is the well-known class of nonlinear systems affine in the control vector. This class of systems includes some important applications, like rigid and flexible robot manipulators (Spong, 1992), some aircraft models (Buffington, 1999), and so on. We focus in the next section on this class of models and propose a constructive method to enforce the ISS property by nonlinear state feedback.

4.5 THE CASE OF NONLINEAR MODELS AFFINE IN THE CONTROL

We consider here affine uncertain nonlinear systems of the form

$$\dot{x} = f(x) + \Delta f(t, x) + g(x)u,$$
$$y = h(x),$$

(4.17)

where $x \in \mathbb{R}^n$, $u \in \mathbb{R}^{na}$, $y \in \mathbb{R}^m$ ($na \geq m$), represent the state, the input, and the controlled output vectors, respectively; $\Delta f(t, x)$ is a vector field representing additive model uncertainties. The vector fields f, Δf, columns of g, and function h satisfy the following assumptions.

   Assumption 4.6. *The function $f: \mathbb{R}^n \to \mathbb{R}^n$ and the columns of $g: \mathbb{R}^n \to \mathbb{R}^{na}$ are C^∞ vector fields on a bounded set X of $\mathbb{R}^n$ and $h: \mathbb{R}^n \to \mathbb{R}^m$ is a C^∞ vector on X. The vector field $\Delta f(x)$ is C^1 on X.*

Assumption 4.7. *System (4.17) has a well-defined (vector) relative degree* $\{r_1, \ldots, r_m\}$ *at each point* $x^0 \in X$, *and the system is linearizable, that is,* $\sum_{i=1}^{m} r_i = n$.

Assumption 4.8. *The desired output trajectories* y_{id} *(1 $\leq$ i $\leq$ m) are smooth functions of time, relating desired initial points* $y_{id}(0)$ *to desired final points* $y_{id}(t_f)$.

4.5.1 Control Objectives

Our objective is to design a state feedback adaptive controller such that the output tracking error is uniformly bounded, whereas the tracking error upper bound is a function of the uncertain parameters' estimation error, which can be decreased by the MES learning. We underline here that the goal of the MES controller is not stabilization, but rather performance optimization; that is, the MES improves the parameters' estimation error, which in turn improves the output tracking error. To achieve this control objective we proceed as follows: First, we design a robust controller which can guarantee input-to-state stability of the tracking error dynamics with respect to the estimation error input. Then we combine this controller with a model-free extremum-seeking algorithm to iteratively estimate the uncertain parameters, by optimizing online a desired learning cost function.

4.5.2 Adaptive Controller Design

The adaptive controller is designed following a two–step procedure. In the first step, we design a nominal controller based on the nominal model of the system, that is, by assuming that the uncertain part is null. Then the uncertainties are added to the model, and the nominal controller is extended to cover the effect of uncertainties by using a Lyapunov reconstruction technique. These two steps are presented next.

4.5.2.1 Nominal Controller

Let us first consider the system under nominal conditions, that is, when $\Delta f(t, x) = 0$. In this case it is well known, see Isidori (1989), that system (4.17) can be written as

$$y^{(r)}(t) = b(\xi(t)) + A(\xi(t))u(t), \tag{4.18}$$

where

$$y^{(r)}(t) = \left[y_1^{(r_1)}(t), y_2^{(r_2)}(t), \dots, y_m^{(r_m)}(t) \right]^T,$$

$$\xi(t) = \left[\xi^1(t), \dots, \xi^m(t) \right]^T, \tag{4.19}$$

$$\xi^i(t) = \left[y_i(t), \dots, y_i^{(r_i-1)}(t) \right], \quad 1 \le i \le m.$$

The functions $b(\xi)$, $A(\xi)$ can be written as functions of f, g, and h, and $A(\xi)$ is nonsingular in $\tilde{X}$, where $\tilde{X}$ is the image of the set of X by the diffeomorphism $x \to \xi$ between the states of system (4.17) and the linearized model (4.18). To deal with the uncertain model we first need to introduce one more assumption on the system (4.17).

Assumption 4.9. *The additive uncertainties $\Delta f(t, x)$ in Eq. (4.17) appear as additive uncertainties in the input-output linearized model (4.18), (4.19) as follows (see also Benosman et al., 2009):*

$$y^{(r)}(t) = b(\xi(t)) + A(\xi(t))u(t) + \Delta b(t, \xi(t)), \tag{4.20}$$

where $\Delta b(t, \xi)$ is C^1 with respect to the state vector $\xi \in \tilde{X}$.

Remark 4.7. *Assumption 4.9 can be ensured under the so-called matching conditions (Elmali and Olgac, 1992, p. 146).*

It is well known that the nominal model (4.18) can be easily transformed into a linear input-output mapping. Indeed, we can first define a virtual input vector $v(t)$ as

$$v(t) = b(\xi(t)) + A(\xi(t))u(t). \tag{4.21}$$

Combining Eqs. (4.18), (4.21) we can obtain the following input-output mapping:

$$y^{(r)}(t) = v(t). \tag{4.22}$$

Based on the linear system (4.22) it is straightforward to design a stabilizing controller for the nominal system (4.18) as

$$u_n = A^{-1}(\xi) \left[v_s(t, \xi) - b(\xi) \right], \tag{4.23}$$

where v_s is an $m \times 1$ vector and the ith ($1 \le i \le m$) element v_{si} is given by

$$v_{si} = y_{id}^{(r_i)} - K_{r_i}^i \left(y_i^{(r_i-1)} - y_{id}^{(r_i-1)} \right) - \cdots - K_1^i (y_i - y_{id}). \tag{4.24}$$

If we denote the tracking error as $e_i(t) \triangleq y_i(t) - y_{id}(t)$, we obtain the following tracking error dynamics:

$$e_i^{(r_i)}(t) + K_{r_i}^i e^{(r_i-1)}(t) + \cdots + K_1^i e_i(t) = 0, \qquad (4.25)$$

where $i \in \{1, 2, \ldots, m\}$. By properly selecting the gains K_j^i where $i \in \{1, 2, \ldots, m\}$ and $j \in \{1, 2, \ldots, r_i\}$, we can obtain global asymptotic stability of the tracking errors $e_i(t)$. To formalize this condition we add the following assumption.

Assumption 4.10. *There exists a nonempty set $\mathcal{A}$ where $K_j^i \in \mathcal{A}$ such that the polynomials in Eq. (4.25) are Hurwitz, where $i \in \{1, 2, \ldots, m\}$ and $j \in \{1, 2, \ldots, r_i\}$.*

To this end, we define $z = [z^1, z^2, \ldots, z^m]^T$, where $z^i = \left[e_i, \dot{e}_i, \ldots, e_i^{(r_i-1)} \right]$, and $i \in \{1, 2, \ldots, m\}$.

Then, from Eq. (4.25), we can obtain

$$\dot{z} = \tilde{A}z,$$

where $\tilde{A} \in \mathbb{R}^{n \times n}$ is a diagonal block matrix given by

$$\tilde{A} = \text{diag}\{\tilde{A}_1, \tilde{A}_2, \ldots, \tilde{A}_m\}, \qquad (4.26)$$

and each $\tilde{A}_i$ $(1 \leq i \leq m)$ is an $r_i \times r_i$ matrix given by

$$\tilde{A}_i = \begin{bmatrix} 0 & 1 & & & \\ 0 & & 1 & & \\ 0 & & & \ddots & \\ \vdots & & & & 1 \\ -K_1^i & -K_2^i & \cdots & \cdots & -K_{r_i}^i \end{bmatrix}.$$

As discussed previously, the gains K_j^i can be chosen such that the matrix $\tilde{A}$ is Hurwitz. Thus there exists a positive definite matrix $P > 0$ such that (see, e.g., Khalil, 2002)

$$\tilde{A}^T P + P\tilde{A} = -I. \qquad (4.27)$$

In the next section we use the nominal controller (4.23) to write a robust ISS controller.

4.5.2.2 Lyapunov Reconstruction-Based ISS Controller

We now consider the uncertain model (4.17), that is, when $\Delta f(t, x) \neq 0$. The corresponding exact linearized model is given by Eq. (4.20) where $\Delta b(t, \xi(t)) \neq 0$. The global asymptotic stability of the error dynamics (4.25) cannot be guaranteed anymore due to the additive uncertainty $\Delta b(t, \xi(t))$.

We use Lyapunov reconstruction techniques to design a new controller so that the tracking error is guaranteed to be bounded given that the estimate error of $\Delta b(t, \xi(t))$ is bounded.

The new controller for the uncertain model (4.20) is defined as

$$u_f = u_n + u_r, \tag{4.28}$$

where the nominal controller u_n is given by Eq. (4.23) and the robust controller u_r will be given later. By using (4.20), (4.23), and (4.28) we obtain

$$
\begin{aligned}
y^{(r)}(t) &= b(\xi(t)) + A(\xi(t))u_f + \Delta b(t, \xi(t)), \\
&= b(\xi(t)) + A(\xi(t))u_n + A(\xi(t))u_r + \Delta b(t, \xi(t)), \\
&= v_s(t, \xi) + A(\xi(t))u_r + \Delta b(t, \xi(t)),
\end{aligned} \tag{4.29}
$$

Which leads to the following error dynamics

$$\dot{z} = \tilde{A}z + \tilde{B}\delta, \tag{4.30}$$

where $\tilde{A}$ is defined in Eq. (4.26), δ is an $m \times 1$ vector given by

$$\delta = A(\xi(t))u_r + \Delta b(t, \xi(t)), \tag{4.31}$$

and the matrix $\tilde{B} \in \mathbb{R}^{n \times m}$ is given by

$$
\tilde{B} = \begin{bmatrix} \tilde{B}_1 \\ \tilde{B}_2 \\ \vdots \\ \tilde{B}_m \end{bmatrix}, \tag{4.32}
$$

where each $\tilde{B}_i (1 \leq i \leq m)$ is given by an $r_i \times m$ matrix such that

$$
\tilde{B}_i(l, q) = \begin{cases} 1 & \text{for } l = r_i, \ q = i, \\ 0 & \text{otherwise.} \end{cases}
$$

If we choose $V(z) = z^T P z$ as a Lyapunov function for the dynamics (4.30), where P is the solution of the Lyapunov equation (4.27), we obtain

$$
\begin{aligned}
\dot{V}(t) &= \frac{\partial V}{\partial z}\dot{z}, \\
&= z^T(\tilde{A}^T P + P\tilde{A})z + 2z^T P\tilde{B}\delta, \\
&= -\|z\|^2 + 2z^T P\tilde{B}\delta,
\end{aligned} \tag{4.33}
$$

where δ given by Eq. (4.31) depends on the robust controller u_r.

Next, we design the controller u_r based on the form of the uncertainties $\Delta b(t, \xi(t))$. More specifically, we consider here the case when $\Delta b(t, \xi(t))$ is of the following form:

$$\Delta b(t, \xi(t)) = EQ(\xi, t), \qquad (4.34)$$

where $E \in \mathbb{R}^{m \times m}$ is a matrix of unknown constant parameters, and $Q(\xi, t): \mathbb{R}^n \times \mathbb{R} \to \mathbb{R}^m$ is a known bounded function of states and time variables. For notational convenience, we denote by $\hat{E}(t)$ the estimate of E, and by $e_E = E - \hat{E}$, the estimate error.

We define the unknown parameter vector $\Delta = [E(1, 1), \ldots, E(m, m)]^T \in \mathbb{R}^{m^2}$, that is, concatenation of all elements of E; its estimate is denoted by $\hat{\Delta}(t) = [\hat{E}(1, 1), \ldots, \hat{E}(m, m)]^T$, and the estimation error vector is given by $e_\Delta(t) = \Delta - \hat{\Delta}(t)$.

Next, we propose the following robust controller:

$$u_r = - A^{-1}(\xi)[\tilde{B}^T Pz \| Q(\xi, t) \|^2 + \hat{E}(t) Q(\xi, t)]. \qquad (4.35)$$

The closed-loop error dynamics can be written as

$$\dot{z} = f(t, z, e_\Delta), \qquad (4.36)$$

where $e_\Delta(t)$ is considered to be an input to the system (4.36).

Theorem 4.4. *Consider the system (4.17), under Assumptions 4.6–4.10, with $\Delta b(t, \xi(t))$ satisfying Eq. (4.34), under the feedback controller (4.28), where u_n is given by Eq. (4.23) and u_r is given by Eq. (4.35). Then, the closed-loop system (4.36) is ISS from the estimation errors input $e_\Delta(t) \in \mathbb{R}^{m^2}$ to the tracking errors state $z(t) \in \mathbb{R}^n$.*

Proof. By substituting Eq. (4.35) into Eq. (4.31), we obtain

$$\delta = - \tilde{B}^T Pz \| Q(\xi, t) \|^2 - \hat{E}(t) Q(\xi, t) + \Delta b(t, \xi(t)),$$

$$= - \tilde{B}^T Pz \| Q(\xi, t) \|^2 - \hat{E}(t) Q(\xi, t) + E Q(\xi, t),$$

if we consider $V(z) = z^T Pz$ as a Lyapunov function for the error dynamics (4.30). Then, from Eq. (4.33), we obtain

$$\dot{V} \le - \|z\|^2 + 2z^T P\tilde{B}EQ(\xi, t) - 2z^T P\tilde{B}\hat{E}(t)Q(\xi, t) - 2\|z^T P\tilde{B}\|^2 \|Q(\xi, t)\|^2,$$

which leads to

$$\dot{V} \le - \|z\|^2 + 2z^T P\tilde{B}e_E Q(\xi, t) - 2\|z^T P\tilde{B}\|^2 \|Q(\xi, t)\|^2.$$

Because

$$z^T P\tilde{B} e_E Q(\xi) \le \|z^T P\tilde{B} e_E Q(\xi)\| \le \|z^T P\tilde{B}\| \|e_E\|_F \|Q(\xi)\|$$
$$= \|z^T P\tilde{B}\| \|e_\Delta\| \|Q(\xi)\|,$$

we obtain

$$\dot{V} \le -\|z\|^2 + 2\|z^T P\tilde{B}\| \|e_\Delta\| \|Q(\xi, t)\| - 2\|z^T P\tilde{B}\|^2 \|Q(\xi, t)\|^2,$$

$$\le -\|z\|^2 - 2\left(\|z^T P\tilde{B}\| \|Q(\xi, t)\| - \frac{1}{2}\|e_\Delta\|\right)^2 + \frac{1}{2}\|e_\Delta\|^2,$$

$$\le -\|z\|^2 + \frac{1}{2}\|e_\Delta\|^2.$$

Thus we have the following relation:

$$\dot{V} \le -\frac{1}{2}\|z\|^2, \quad \forall \|z\| \ge \|e_\Delta\| > 0.$$

Then from Eq. (4.2), we obtain that system (4.36) is ISS from input e_Δ to state z.

Remark 4.8. *In the case of constant uncertainty vector, that is, $\Delta b = \Delta = cte \in \mathbb{R}^m$, the controller (4.35) boils down to the simple feedback*

$$u_r = -A^{-1}(\xi)[\tilde{B}^T Pz + \hat{\Delta}(t)]. \tag{4.37}$$

4.5.2.3 MES-Based Parametric Uncertainties Estimation

Let us define now the following cost function:

$$J(\hat{\Delta}) = F(z(\hat{\Delta})), \tag{4.38}$$

where $F: \mathbb{R}^n \to \mathbb{R}$, $F(0) = 0$, $F(z) > 0$ for $z \in \mathbb{R}^n - \{0\}$. We need the following assumptions on J.

Assumption 4.11. *The cost function J has a local minimum at $\hat{\Delta}^* = \Delta$.*

Assumption 4.12. *The initial error $e_\Delta(t_0)$ is sufficiently small; that is, the original parameter estimate vector $\hat{\Delta}$ is close enough to the actual parameter vector Δ.*

Assumption 4.13. *The cost function J is analytic and its variation with respect to the uncertain parameters is bounded in the neighborhood of $\hat{\Delta}^*$, that is, $\|\frac{\partial J}{\partial \hat{\Delta}}(\tilde{\Delta})\| \le \xi_2$, $\xi_2 > 0$, $\tilde{\Delta} \in \mathcal{V}(\hat{\Delta}^*)$, where $\mathcal{V}(\hat{\Delta}^*)$ denotes a compact neighborhood of $\hat{\Delta}^*$.*

Remark 4.9. *Assumption 4.11 simply states that the cost function J has at least a local minimum at the true values of the uncertain parameters.*

Remark 4.10. *Assumption 4.12 indicates that our results are of local nature; that is, our analysis holds in a small neighborhood of the actual values of the uncertain parameters. This is usually the case in real applications when the adaptation of the controller is meant for tracking a slow aging of a system, that is, slow drifts of some of its parameters' nominal values.*

We can now present the following result.

Lemma 4.1. *Consider the system (4.20), (4.34) with the cost function (4.38), under Assumptions 4.6–4.13, with the feedback controller (4.28), where u_n is given by Eq. (4.23), u_r is given by Eq. (4.35), and $\hat{\Delta}(t)$ are estimated through the MES algorithm*

$$\dot{x}_{\Delta i} = a_i \sin\left(\omega_i t + \frac{\pi}{2}\right) J(\hat{\Delta}),$$

$$\hat{\Delta}_i(t) = x_{\Delta i} + a_i \sin\left(\omega_i t - \frac{\pi}{2}\right), \quad i \in \{1, 2, \dots, m^2\} \qquad (4.39)$$

with $\omega_i \neq \omega_j$, $\omega_i + \omega_j \neq \omega_k$, $i, j, k \in \{1, 2, \dots, m^2\}$, and $\omega_i > \omega^*$, $\forall\ i \in \{1, 2, \dots, m^2\}$, with ω^* large enough. Then, the norm of the error vector $z(t)$ admits the following bound:

$$\|z(t)\| \leq \beta(\|z(0)\|, t) + \gamma(\tilde{\beta}(\|e_\Delta(0)\|, t) + \|e_\Delta\|_{\max}),$$

where $\|e_\Delta\|_{\max} = \frac{\xi_1}{\omega_0} + \sqrt{\sum_{i=1}^{m^2} a_i^2}$, $\xi_1 > 0$, $e_\Delta(0) \in \mathcal{D}_e$, $\omega_0 = \max_{i \in \{1,2,\dots,m^2\}} \omega_i$, $\beta \in \mathcal{KL}$, $\tilde{\beta} \in \mathcal{KL}$, and $\gamma \in \mathcal{K}$.

Proof. Based on Theorem 4.4, we know that the tracking error dynamics (4.36) is ISS from the input $e_\Delta(t)$ to the state $z(t)$. Thus, by Definition 4.1, there exist a class $\mathcal{KL}$ function β and a class $\mathcal{K}$ function γ such that for any initial state $z(0)$, any bounded input $e_\Delta(t)$, and any $t \geq 0$, we have

$$\|z(t)\| \leq \beta(\|z(0)\|, t) + \gamma\left(\sup_{0 \leq \tau \leq t} \|e_\Delta(\tau)\|\right). \qquad (4.40)$$

Now, we need to evaluate the bound on the estimation vector $\hat{\Delta}(t)$; to do so we use the results presented in Rotea (2000). First, based on Assumption 4.13, the cost function is locally Lipschitz; that is, there exists $\eta_1 > 0$ such that $|J(\Delta_1) - J(\Delta_2)| \leq \eta_1 \|\Delta_1 - \Delta_2\|$, for all $\Delta_1, \Delta_2 \in \mathcal{V}(\hat{\Delta}^*)$. Furthermore, because J is analytic, it can be approximated locally in $\mathcal{V}(\hat{\Delta}^*)$ by a quadratic function, for example, Taylor series up to the second order. Based on this and on Assumptions 4.11 and 4.12, we can obtain the following bound (Rotea, 2000; Benosman and Atinc, 2015b):

$$\|e_\Delta(t)\| - \|d(t)\| \leq \|e_\Delta(t) - d(t)\| \leq \tilde{\beta}(\|e_\Delta(0)\|, t\|) + \frac{\xi_1}{\omega_0},$$

where $\tilde{\beta} \in \mathcal{KL}$, $\xi_1 > 0$, $t \geq 0$, $\omega_0 = \max_{i \in \{1,2,\dots,m^2\}} \omega_i$, and $d(t) = [a_1 \sin(\omega_1 t + \frac{\pi}{2}), \dots, a_{m^2} \sin(\omega_{m^2} t + \frac{\pi}{2})]^T$.

We can further write that

$$\|e_\Delta(t)\| \leq \tilde{\beta}(\|e_\Delta(0)\|, t\|) + \frac{\xi_1}{\omega_0} + \|d(t)\|,$$

$$\leq \tilde{\beta}(\|e_\Delta(0)\|, t\|) + \frac{\xi_1}{\omega_0} + \sqrt{\sum_{i=1}^{m^2} a_i^2},$$

which together with Eq. (4.40) leads to the desired result.

Remark 4.11. *The adaptive controller of Lemma 4.1 uses the ES algorithm (4.39) to estimate the model parametric uncertainties. One might wonder where the persistence of excitation (PE) condition appears (please refer to Chapter 1 for more details about the PE condition). The answer can be found in the examination of Eq. (4.39). Indeed, the ES algorithm uses as input the sinusoidal signals $a_i \sin(\omega_i t + \frac{\pi}{2})$, which clearly satisfy the PE condition. The main difference with some classical adaptive control results is that these excitation signals are not entering the system dynamics directly, but instead are applied as input to the ES algorithm, then reflected on the ES estimation outputs, and thus transmitted to the system through the feedback loop.*

Remark 4.12. *One main difference with some of the existing model-based adaptive controllers is the fact that the ES estimation algorithm does not depend on the model of the system; that is, the only information needed to compute the learning cost function (4.38) is the desired trajectory and the measured output of the system (please refer to Section 4.6 for an example). This makes the ES-based adaptive controllers suitable for the general case of nonlinear parametric uncertainties. For example, in Benosman (2013), a similar algorithm has been tested in the case of nonlinear models of electromagnetic actuators with a nonlinear parametric uncertainty. Another point worth mentioning here is the fact that with the available model-based adaptive controllers like the X-swapping modular algorithms, see for example, Krstic et al. (1995), it is not possible to estimate multiple uncertainties simultaneously in some cases. For instance, it is shown in Benosman and Atinc (2015a) that the X-swapping adaptive control cannot estimate multiple uncertainties in the case of electromagnetic actuators, due to the linear dependency of the uncertain parameters. However, when dealing with the same example, the ES-based adaptive control approach was successful in dealing with multiple uncertainties; see Benosman and Atinc (2015b).*

As we mentioned before, the class of systems considered in Sections 4.3 and 4.4 are quite general. In the next section, we will see how to apply these results to the case of electromagnetic actuators. We will also use the results presented in Section 4.5 to study the case of rigid manipulators, which are well known to be modeled by a nonlinear model affine in the control.

4.6 MECHATRONICS EXAMPLES

4.6.1 The Case of Electromagnetic Actuators

First, let us recall the dynamical model of electromagnetic actuators, see Peterson and Stefanopoulou (2004), given by

$$
\begin{aligned}
m\frac{d^2x}{dt^2} &= k(x_0 - x) - \eta\frac{dx}{dt} - \frac{ai^2}{2(b+x)^2}, \\
u &= Ri + \frac{a}{b+x}\frac{di}{dt} - \frac{ai}{(b+x)^2}\frac{dx}{dt}, \quad 0 \le x \le x_f,
\end{aligned}
\tag{4.41}
$$

where x represents the armature position, which is constrained between the initial position of the armature 0 and the maximal position of the armature x_f, $\frac{dx}{dt}$ represents the armature velocity, m is the armature mass, k the spring constant, x_0 the initial spring length, η the damping coefficient, $\frac{ai^2}{2(b+x)^2}$ represents the electromagnetic force (EMF) generated by the coil, a, b being constant parameters of the coil, R the resistance of the coil, $L = \frac{a}{b+x}$ the coil inductance (assumed to be dependent on the position of the armature), and $\frac{ai}{(b+x)^2}\frac{dx}{dt}$ represents the back EMF. Finally, i denotes the coil current, $\frac{di}{dt}$ its time derivative, and u represents the control voltage applied to the coil.

We consider here the case where the electromagnetic system has an uncertain spring constant k, and an uncertain damping coefficient η.

4.6.1.1 Controller Design

Let us define the state vector $\mathbf{z} := [z_1, z_2, z_3]^T = [x, \dot{x}, i]^T$. The objective of the control is to make the variables (z_1, z_2) robustly track a sufficiently smooth (i.e., at least C^2) time-varying position and velocity trajectories $z_1^{\text{ref}}(t), z_2^{\text{ref}}(t) = \frac{dz_1^{\text{ref}}(t)}{dt}$, which satisfy the following constraints: $z_1^{\text{ref}}(t_0) = z_{1_{\text{int}}}, z_1^{\text{ref}}(t_f) = z_{1_f}, \dot{z}_1^{\text{ref}}(t_0) = \dot{z}_1^{\text{ref}}(t_f) = 0, \ddot{z}_1^{\text{ref}}(t_0) = \ddot{z}_1^{\text{ref}}(t_f) = 0$, where t_0 is the starting time of the trajectory, t_f is the final time, $z_{1_{\text{int}}}$ is the initial position, and z_{1_f} is the final position.

To start, we first write the system (4.41) in the following form:

$$\dot{z}_1 = z_2,$$
$$\dot{z}_2 = \frac{k}{m}(x_0 - z_1) - \frac{\eta}{m}z_2 - \frac{a}{2m(b + z_1)^2}z_3^2,$$
$$\dot{z}_3 = -\frac{R}{\frac{a}{b+z_1}}z_3 + \frac{z_3}{b + z_1}z_2 + \frac{u}{\frac{a}{b+z_1}}.$$

(4.42)

The authors in Benosman and Atinc (2015b) have shown that if we select the following feedback controller

$$u = \frac{a}{b + z_1}\left(\frac{R(b + z_1)}{a}z_3 - \frac{z_2 z_3}{(b + z_1)} + \frac{1}{2z_3}\right.$$

$$\left(\frac{a}{2m(b + z_1)^2}(z_2 - z_2^{\text{ref}}) - c_2(z_3^2 - \tilde{u})\right)\right)$$

$$+ \frac{2mz_2}{z_3}\left(\frac{\hat{k}}{m}(x_0 - z_1) - \frac{\hat{\eta}}{m}z_2 + c_3(z_1 - z_1^{\text{ref}}) + c_1(z_2 - z_2^{\text{ref}}) - \dot{z}_2^{\text{ref}}\right.$$

$$\left. + \kappa_1(z_2 - z_2^{\text{ref}})\|\psi\|_2^2\right) + \frac{m(b + z_1)}{z_3}$$

$$\left(\left(\frac{\hat{k}}{m}(x_0 - z_1) - \frac{\hat{\eta}}{m}z_2 - \frac{a}{2m(b + z_1)^2}z_3^2 - \dot{z}_2^{\text{ref}}\right)\right.$$

$$\left(c_1 + \kappa_1\|\psi\|_2^2 - \frac{\hat{\eta}}{m}\right) - \frac{\hat{\eta}}{m}\dot{z}_2^{\text{ref}}\right) + \frac{m(b + z_1)}{z_3}\left(2\kappa_1(z_2 - z_2^{\text{ref}})\right)$$

$$\times \left(\frac{(x_0 - z_1)(-z_2)}{m^2} + \frac{z_2\left(\frac{\hat{k}}{m}(x_0 - z_1) - \frac{\hat{\eta}}{m}z_2 - \frac{az_3^2}{2m(b+z_1)^2}\right)}{m^2}\right)$$

$$- \kappa_2(z_3^2 - \tilde{u})\left|\frac{m(b + z_1)}{z_3}\right|^2\left[\left|c_1 + \kappa_1\|\psi\|_2^2 - \frac{\hat{\eta}}{m}\right|^2\right.$$

$$\left. + \left|2\kappa_1(z_2 - z_2^{\text{ref}})\right|^2\left|\frac{z_2}{m^2}\right|^2\right]\|\psi\|_2^2$$

$$- \kappa_3(z_3^2 - \tilde{u})\left|\frac{m(b + z_1)}{z_3}\right|^2\|\psi\|_2^2 + \frac{m(b + z_1)}{z_3}$$

$$\left(-\frac{\hat{k}}{m}z_2 - \ddot{z}_2^{\text{ref}} + c_3(z_2 - z_2^{\text{ref}})\right),$$

(4.43)

with

$$
\tilde{u} = \frac{2m(b+z_1)^2}{a} \left(\frac{\hat{k}}{m}(x_0 - z_1) - \frac{\hat{\eta}}{m}z_2 + c_3(z_1 - z_1^{\mathrm{ref}}) \right.
$$

$$
\left. + c_1(z_2 - z_2^{\mathrm{ref}}) - \dot{z}_2^{\mathrm{ref}} \right) + \frac{2m(b+z_1)^2}{a} \left(\kappa_1(z_2 - z_2^{\mathrm{ref}}) \| \psi \|_2^2 \right), \tag{4.44}
$$

where $\hat{k}, \hat{\eta}$ are the parameter estimates, and $\psi \triangleq (\frac{x_0 - z_1}{m}, \frac{z_2}{m}, \frac{1}{m})^T$, then the system (4.41), (4.43), and (4.44) is LiISS.

Next, we define the cost function

$$
Q(\hat{\Delta}) = \int_0^{t_f} Q_1(z_1(s) - z_1(s)^{\mathrm{ref}})^2 ds + \int_0^{t_f} Q_2(z_2(s) - z_2^{\mathrm{ref}}(s))^2 ds, \tag{4.45}
$$

where $Q_1; Q_2 > 0$, and $\hat{\Delta} = (\hat{\Delta}_k, \hat{\Delta}_\eta)^T$ represents the vector of the learned parameters, defined such that

$$
\hat{k}(t) = k_{\mathrm{nominal}} + \hat{\Delta}_k(t),
$$
$$
\hat{\eta}(t) = \eta_{\mathrm{nominal}} + \hat{\Delta}_\eta(t), \tag{4.46}
$$

where $k_{\mathrm{nominal}}, \eta_{\mathrm{nominal}}$ are the nominal values of the parameters, and the $\hat{\Delta}_i$s are computed using a discrete version of Eq. (4.12), given by

$$
x_k(k'+1) = x_k(k') + a_k t_f \sin \left(\omega_k k' t_f + \frac{\pi}{2} \right) Q,
$$

$$
\hat{\Delta}_k(k'+1) = x_k(k'+1) + a_k \sin \left(\omega_k k' t_f - \frac{\pi}{2} \right),
$$

$$
x_\eta(k'+1) = x_\eta(k') + a_\eta t_f \sin \left(\omega_\eta k' t_f + \frac{\pi}{2} \right) Q, \tag{4.47}
$$

$$
\hat{\Delta}_\eta(k'+1) = x_\eta(k'+1) + a_\eta \sin \left(\omega_\eta k' t_f - \frac{\pi}{2} \right).
$$

Eventually, we conclude based on Theorem 4.2 that the controller (4.43), (4.44), (4.46), and (4.47) ensures that the norm of the tracking error is bounded with a decreasing function of the estimation error.

4.6.1.2 Numerical Results

We illustrate our approach for the nonlinear electromagnetic actuator modeled by Eq. (4.41) using the system parameters given in Table 4.1. The reference trajectory is designed to be a fifth-order polynomial

Table 4.1 Numerical values of the
mechanical parameters for the
electromagnetic example

Parameter	Value
m	0.3 kg
R	6.5 Ω
η	8 kg/s
x_0	8 mm
k	160 N/mm
a	15×10^{-6} Nm^2/A^2
b	4.5×10^{-5} m

$x^{\text{ref}}(t) = \sum_{i=0}^{5} a_i(\frac{t}{t_f})^i$, where the coefficients a_i are selected such that the following conditions are satisfied: $x^{\text{ref}}(0) = 0.1$ mm, $x^{\text{ref}}(0.5) = 0.6$ mm, $\dot{x}^{\text{ref}}(0) = 0$, $\dot{x}^{\text{ref}}(0.5) = 0$, $\ddot{x}^{\text{ref}}(0) = 0$, $\ddot{x}^{\text{ref}}(0.5) = 0$.

We consider the uncertainties given by $\Delta k = -4$, and $\Delta \eta = -0.8$. We implement the controller (4.43), (4.44) with the coefficients $c_1 = 100, c_2 = 100, c_3 = 2500, \kappa_1 = \kappa_2 = \kappa_3 = 0.25$, together with the learning algorithm (4.45), (4.46), and (4.47), with the coefficients $a_k = 0.3$, $\omega_k = 8$ rad/s, $a_\eta = 0.1$, $\omega_\eta = 8.4$ rad/s, $Q_1 = Q_2 = 100$. For more details about the tuning of the MES coefficients, we refer the reader to Rotea (2000) and Ariyur and Krstic (2002, 2003). However, we underline here that the frequencies $\omega_i, 1 = 1, 2$ are selected high enough to ensure efficient exploration on the search space and ensure convergence. The amplitudes $a_i, i = 1, 2$ of the dither signals are chosen such that the search is fast enough for this application. Here, due to the cyclic nature of the problem, that is, cyclic motion of the armature between 0 and x_f, the uncertain parameters' estimate vector $(\hat{k}, \hat{\eta})$ is updated for each cycle. Indeed, at the end of each cycle at $t = t_f$ the cost function Q is updated, and the new estimate of the parameters is computed for the next cycle. The purpose of using the MES scheme along with the LiISS backstepping controller is to improve the performance of the LiISS backstepping controller by improving the estimate of the system's parameters over many cycles, hence decreasing the tracking error over time.

As it can be seen in Fig. 4.1A and B, the robustification of the backstepping control via ES greatly improves the tracking performance. Fig. 4.2A shows that the cost function decreases below 2 within 20 iterations.

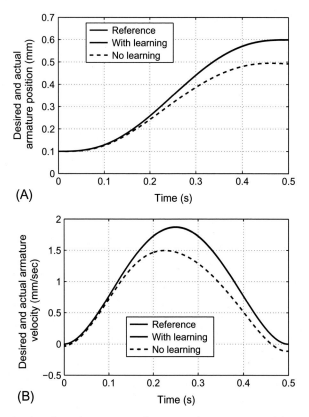

(A)

(B)

Fig. 4.1 Obtained trajectories versus reference trajectory: case with uncertain k, η. (A) Obtained armature position versus reference trajectory. (B) Obtained armature velocity versus reference trajectory.

It can be seen in Fig. 4.2B that the cost starts at an initial value around 12, and decreases rapidly afterwards. Moreover, the estimated parametric uncertainties $\hat{\Delta} k$, and $\hat{\Delta} \eta$ converge to regions around the actual parameters' values, as shown in Fig. 4.3. The number of iterations for the estimate to converge may appear to be high. The reason behind that is that the allowed uncertainties in the parameters are large. Thus the extremum-seeking scheme requires a lot of iterations to improve the performance. Furthermore, we purposely tested the challenging case of two simultaneous uncertainties, which makes the space search for the learning algorithm large. Note that this case of multiple uncertainties could not be solved with other classical model-based adaptive controllers, see Benosman and Atinc

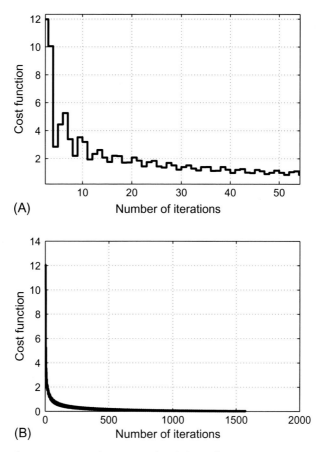

Fig. 4.2 Cost function: case with uncertain k, η. (A) Cost function: zoom. (B) Cost function.

(2013b), due to some intrinsic limitations of classical model-based adaptive controllers (please refer to Chapter 2 for more details about the difference between learning-based and model-based adaptive control). However, in real-life applications it is often the case that uncertainties accumulate gradually over a long period of time, while the learning algorithm keeps tracking these changes continuously. Thus the extremum-seeking algorithm will be able to improve the controller performance in much fewer iterations.

Next, we want to test the more challenging case of nonlinear uncertainty. Namely, we consider the case where the coefficient b in the model (4.41) is uncertain, and try to estimate online the true

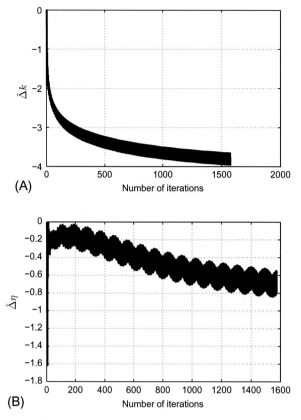

Fig. 4.3 Parameters' estimates: case with uncertain k, η. (A) Parameter k estimate. (B) Parameter η estimate.

value of this parameter. We assume that the value of the uncertainty is $\Delta b = -0.005$. Then, the controller (4.43), (4.44) is applied to the system together with the ES estimator

$$x_b(k' + 1) = x_b(k') + a_b t_f \sin\left(\omega_b k' t_f + \frac{\pi}{2}\right) Q,$$

$$\hat{\Delta}_b(k' + 1) = x_b(k' + 1) + a_b \sin\left(\omega_b k' t_f - \frac{\pi}{2}\right), \quad k' = 0, 1, 2, \ldots$$

$$(4.48)$$

where the cost function Q is given by Eq. (4.45). Similar to the previous case, the estimate of the uncertain parameter $\hat{b}$ is given by

$$\hat{b}(t) = b_{\text{nominal}} + \hat{\Delta}_b(t). \tag{4.49}$$

We simulate the controller, (4.43), (4.44), (4.45), (4.48), and (4.49), with the following parameters $c_1 = 100, c_2 = 100, c_3 = 2500, \kappa_1 = \kappa_2 = \kappa_3 = 0.25$, $Q_1 = Q_2 = 100$, $a_b = 10^{-3}$, $\omega_b = 7.6$ rad/s. The obtained results are reported in Fig. 4.4. One can see clearly in Fig. 4.4A that the cost function starts at a value of 5 and then decreases very quickly in few iterations (less than 5 iterations). The estimation of b is shown in Fig. 4.4B, where we can see a convergence to the actual value of the parameter. Finally, the tracking performance of the desired armature position and velocity are reported in Fig. 4.4C and D, where it is clear that the performance improves greatly with learning; that is, when the learning is applied, we cannot even distinguish between the reference and the actual trajectories.

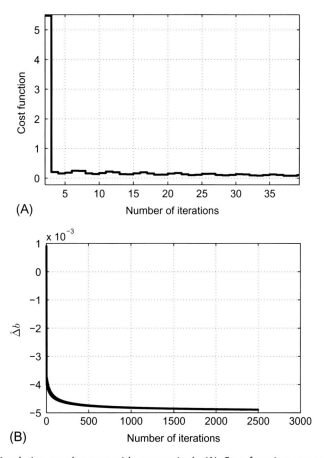

Fig. 4.4 Simulation results: test with uncertain b. (A) Cost function versus learning iterations. (B) $\hat{\Delta}_b$ versus learning iterations.

(Continued)

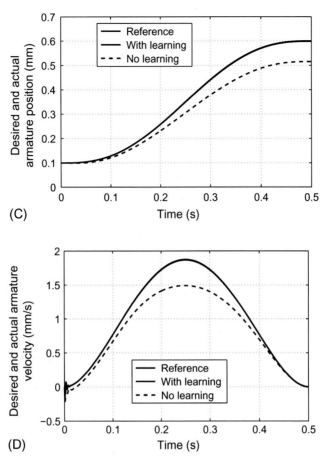

Fig. 4.4 Cont'd, (C) Obtained armature position versus reference trajectory. (D) Obtained armature velocity versus reference trajectory.

4.6.2 The Case of Two-Link Rigid Manipulators

We consider here a two-link robot manipulator, with the following dynamics:

$$H(q)\ddot{q} + C(q, \dot{q})\dot{q} + G(q) = \tau, \tag{4.50}$$

where $q \triangleq [q_1, q_2]^T$ denotes the two joint angles, and $\tau \triangleq [\tau_1, \tau_2]^T$ denotes the two joint torques. The matrix H is assumed to be nonsingular and is given by

$$H \triangleq \begin{bmatrix} H_{11} & H_{12} \\ H_{21} & H_{22} \end{bmatrix},$$

where

$$H_{11} = m_1\ell_{c_1}^2 + I_1 + m_2[\ell_1^2 + \ell_{c_2}^2 + 2\ell_1\ell_{c_2}\cos(q_2)] + I_2,$$
$$H_{12} = m_2\ell_1\ell_{c_2}\cos(q_2) + m_2\ell_{c_2}^2 + I_2,$$
$$H_{21} = H_{12},$$
$$H_{22} = m_2\ell_{c_2}^2 + I_2.$$

(4.51)

The matrix $C(q, \dot{q})$ is given by

$$C(q, \dot{q}) \triangleq \begin{bmatrix} -h\dot{q}_2 & -h\dot{q}_1 - h\dot{q}_2 \\ h\dot{q}_1 & 0 \end{bmatrix},$$

where $h = m_2\ell_1\ell_{c_2}\sin(q_2)$.

The vector $G = [G_1, G_2]^T$ is given by

$$G_1 = m_1\ell_{c_1}g\cos(q_1) + m_2g[\ell_2\cos(q_1 + q_2) + \ell_1\cos(q_1)],$$
$$G_2 = m_2\ell_{c_2}g\cos(q_1 + q_2),$$

(4.52)

where ℓ_1, ℓ_2 are the lengths of the first and second link, respectively, and ℓ_{c_1}, ℓ_{c_2} are the distances between the rotation center and the center of mass of the first and second link, respectively. m_1, m_2 are the masses of the first and second link, respectively; I_1 is the moment of inertia of the first link and I_2 the moment of inertia of the second link, respectively; and g denotes the earth gravitational constant.

In our simulations, we assume that the parameters take the values summarized in Table 4.2. The system dynamics (4.50) can be rewritten as

$$\ddot{q} = H^{-1}(q)\tau - H^{-1}(q)[C(q, \dot{q})\dot{q} + G(q)].$$

(4.53)

Table 4.2 System parameters for the manipulator example

Parameter	Value
I_2	$\frac{5.5}{12}$ kg m^2
m_1	10.5 kg
m_2	5.5 kg
ℓ_1	1.1 m
ℓ_2	1.1 m
ℓ_{c_1}	0.5 m
ℓ_{c_2}	0.5 m
I_1	$\frac{11}{12}$ kg m^2
g	9.8 m/s^2

Thus the nominal controller is given by

$$\tau_n = [C(q,\dot{q})\dot{q} + G(q)] + H(q)\left[\ddot{q}_d - K_d(\dot{q} - \dot{q}_d) - K_p(q - q_d)\right], \quad (4.54)$$

where $q_d = [q_{1d}, q_{2d}]^T$ denotes the desired trajectory, and the diagonal gain matrices $K_p > 0$, $K_d > 0$ are chosen such that the linear error dynamics (as in Eq. 4.25) are asymptotically stable. We choose as output references the fifth-order polynomials $q_{1ref}(t) = q_{2ref}(t) = \sum_{i=0}^{5} a_i(t/t_f)^i$, where the a_i's have been computed to satisfy the boundary constraints $q_{iref}(0) = 0, q_{iref}(t_f) = q_f, \dot{q}_{iref}(0) = \dot{q}_{iref}(t_f) = 0, \ddot{q}_{iref}(0) = \ddot{q}_{iref}(t_f) = 0$, $i = 1, 2$, with $t_f = 2$ s, $q_f = 1.5$ rad. In these tests we assume that the nonlinear model (4.50) is uncertain. In particular, we assume that there exist additive uncertainties in the model (4.53); that is,

$$\ddot{q} = H^{-1}(q)\tau - H^{-1}(q)\left[C(q,\dot{q})\dot{q} + G(q)\right] - EG(q), \quad (4.55)$$

where E is a matrix of constant uncertain parameters. Following Eq. (4.35) the robust part of the control writes as

$$\tau_r = -H(\tilde{B}^T Pz\|G\|^2 - \hat{E}G(q)), \quad (4.56)$$

where

$$\tilde{B}^T = \begin{bmatrix} 0 & 1 & 0 & 0 \\ 0 & 0 & 0 & 1 \end{bmatrix},$$

P is solution of the Lyapunov equation (4.27), with

$$\tilde{A} = \begin{bmatrix} 0 & 1 & 0 & 0 \\ -K_p^1 & -K_d^1 & 0 & 0 \\ 0 & 0 & 0 & 1 \\ 0 & 0 & -K_p^2 & -K_d^2 \end{bmatrix},$$

$z = [q_1 - q_{1d}, \dot{q}_1 - \dot{q}_{1d}, q_2 - q_{2d}, \dot{q}_2 - \dot{q}_{2d}]^T$, and $\hat{E}$ is the matrix of the parameters' estimates. Eventually, the final feedback controller writes as

$$\tau = \tau_n + \tau_r. \quad (4.57)$$

4.6.3 MES-Based Uncertainties Estimation

We consider here two cases: the case where E is diagonal, and the case where E has two uncertain parameters appearing on the same line, that is, the case where the uncertain parameters are linearly dependent. The last case is more challenging than the first one because the uncertainties' effect is not observable from the measured output. More explanations are given next, when reporting the obtained results for each case.

Case 1: Diagonal E. We begin with the case when E is a diagonal matrix, where $E(i, i) = \Delta_i, i = 1, 2$. The controller is designed according to Theorem 4.4. We simulate the case where $\Delta_1 = 0.5$ and $\Delta_2 = 0.3$. The estimate of the two parameters $\hat{\Delta}_i$ $(i = 1, 2)$ are computed using a discrete version of Eq. (4.39), given by

$$
\begin{aligned}
x_{\Delta i}(k + 1) &= x_{\Delta i}(k) + a_i t_f \sin\left(\omega_i t_f k + \frac{\pi}{2}\right) J(\hat{\Delta}), \\
\hat{\Delta}_i(k + 1) &= x_{\Delta i}(k + 1) + a_i \sin\left(\omega_i t_f k - \frac{\pi}{2}\right), \quad i = 1, 2
\end{aligned}
\tag{4.58}
$$

where $k \in \mathbb{N}$ denotes the iteration index, and $x_i(0) = \hat{\Delta}_i(0) = 0$. We choose the following learning cost function:

$$
\begin{aligned}
J(\hat{\Delta}) &= \int_0^{t_f} (q(\hat{\Delta}) - q_d(t))^T Q_1 (q(\hat{\Delta}) - q_d(t)) dt \\
&\quad + \int_0^{t_f} (\dot{q}(\hat{\Delta}) - \dot{q}_d(t))^T Q_2 (\dot{q}(\hat{\Delta}) - \dot{q}_d(t)) dt,
\end{aligned}
\tag{4.59}
$$

where $Q_1 > 0$ and $Q_2 > 0$ denote the weight matrices. The parameters used in the MES (4.58) are summarized in Table 4.3. The uncertain parameter vector $\hat{\Delta}$ is updated for each cycle; that is, at the end of each cycle at $t = t_f$ the cost function J is updated, and the new estimate of the parameters is computed for the next cycle. We first report in Fig. 4.5 the learning cost function. It is clear that the cost function decreases over the learning iterations. The corresponding tracking performance is shown in Figs. 4.8–4.11. We see that without the learning part, that is, when applying the controller (4.57) with nominal Δ values, the tracking performance is poor. Whereas after learning the actual values of the uncertainties, as seen in Figs. 4.6 and 4.7, a good tracking performance is recovered.

Case 2: Linearly dependent uncertainties. In the case where the uncertainties enter the model in a linearly dependent function (e.g., when the matrix E has only one nonzero line), some of the classical

Table 4.3 Parameters used in the MES algorithm for Case 1

a_1	a_2	ω_1	ω_2
0.05	0.05	7	5

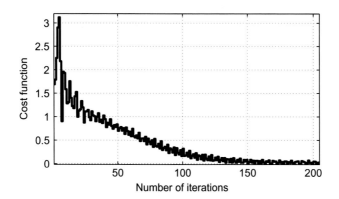

Fig. 4.5 Cost function over the learning iterations (Case 1).

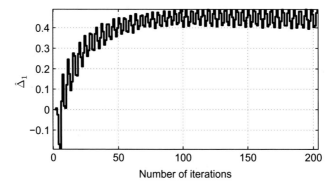

Fig. 4.6 Estimate of Δ_1 over the learning iterations (Case 1).

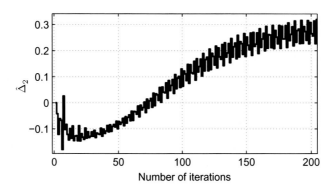

Fig. 4.7 Estimate of Δ_2 over the learning iterations (Case 1).

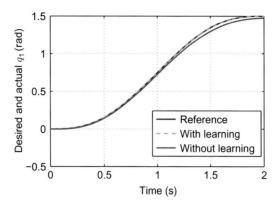

Fig. 4.8 Obtained versus desired first angular trajectory (Case 1).

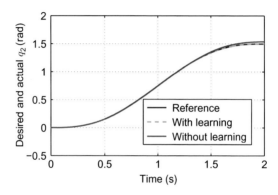

Fig. 4.9 Obtained versus desired second angular trajectory (Case 1).

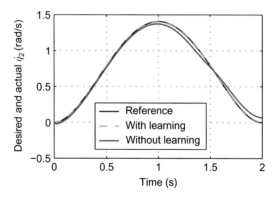

Fig. 4.10 Obtained versus desired first angular velocity trajectory (Case 1).

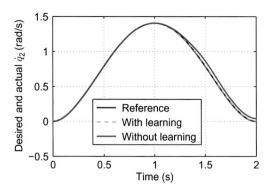

Fig. 4.11 Obtained versus desired second angular velocity trajectory (Case 1).

model-based adaptive controllers (e.g., X-swapping controllers) cannot be used to estimate all the uncertain parameters simultaneously. For example, it has been shown in Benosman and Atinc (2015a) that the model-based gradient descent filters failed to estimate simultaneously multiple parameters in the case of the electromagnetic actuator example.

More specifically, we consider here the following case: $E(1, 1) = 0.3$, $E(1, 2) = 0.6$, and $E(2, i) = 0, i = 1, 2$. In this case, the uncertainties' effect on the acceleration $\ddot{q}_1$ cannot be differentiated, and thus the application of the model-based adaptive method to estimate the actual values of both uncertainties is challenging. However, we show next that by using the controller (4.54), (4.56), (4.57), and (4.58), we manage to estimate the actual values of the uncertainties and improve the tracking performance. Indeed, we implemented the MES-based adaptive controller with the learning parameters shown in Table 4.4. The obtained performance cost function is displayed in Fig. 4.12, where we see that the performance improves over the learning iterations. The corresponding parameters' estimation profiles are reported in Figs. 4.13 and 4.14, which show a quick convergence of the first estimate $\hat{\Delta}_1$ to a neighborhood of the actual value. The convergence of the second estimate $\hat{\Delta}_2$ is slower, which is expected from MES algorithms when many parameters are estimated at the same time. One has to underline here, however, that the convergence speed of the

Table 4.4 Parameters used in MES for Case 2

a_1	a_2	ω_1	ω_2
0.1	0.05	7	5

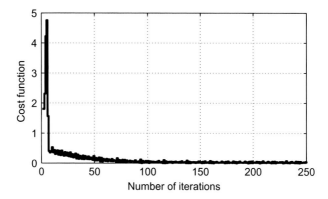

Fig. 4.12 Cost function over the learning iterations (Case 2).

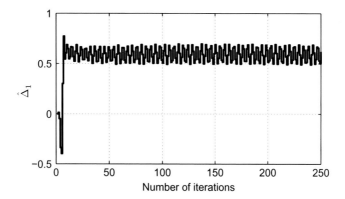

Fig. 4.13 Estimate of Δ_1 over the learning iterations (Case 2).

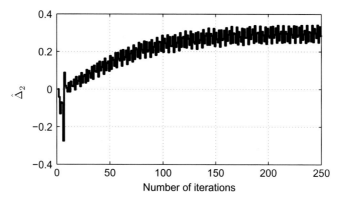

Fig. 4.14 Estimate of Δ_2 over the learning iterations (Case 2).

estimates and the excursion around their final mean values can be directly fine-tuned by the proper choice of the learning coefficients $a_i, \omega_i, i = 1, 2$ in Eq. (4.58). For instance, it is well known, see Moase et al. (2009), that choosing varying coefficients, which start with a high value to accelerate the search initially and then are tuned down when the cost function becomes smaller, accelerates the learning and achieves a convergence to a tighter neighborhood of the local optimum (due to the decrease of the dither amplitudes). Another well-known solution is to introduce an extra tuning parameter, that is, an integrator gain, in the MES algorithm. Finally, the tracking performance is shown in Figs. 4.15 and 4.16, where we can see that after learning the actual values of the uncertainties, the tracking of the desired trajectories is recovered. We only show the first angular trajectories

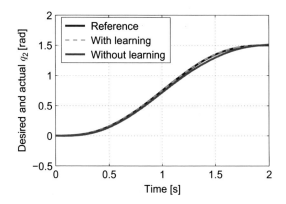

Fig. 4.15 Obtained versus desired first angular trajectory (Case 2).

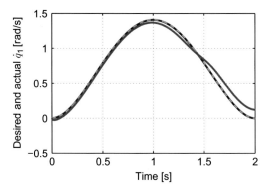

Fig. 4.16 Obtained versus desired first angular velocity trajectory (Case 2).

here because the uncertainties affect only the acceleration $\ddot{q}_1$, and their effect on the tracking for the second angular variable is negligible.

4.7 CONCLUSION

In this chapter we have focused on the problem of controlling systems which are modeled by nonlinear dynamics with uncertain parameters. To control this type of system, we adopted the active approach. Indeed, we did not rely on the worst-case scenario type of approaches which use, for example, interval-based robust control theory to deal with parametric uncertainties. Instead, we focused on developing adaptive controllers that track and estimate online the actual value of the uncertainties. To do so, we chose a modular indirect adaptive control approach. In this method, one has to first design a model-based stabilizing controller based on the best estimate of the uncertainties (or assuming the best estimate, i.e., certainty equivalence paradigm), and then complement this controller with filters or estimators to estimate online the actual values of the uncertainties, and thus improve the overall performance of the adaptive controller. However, we proposed here a new way of realizing a modular adaptive controller, in the sense that the estimation filters that we used are not based on the model of the system; instead, they only rely on some real-time measurements from the system.

We started this chapter by considering a quite general case of nonlinear systems, with basic smoothness assumptions (to guarantee existence and uniqueness of solutions), but without assuming any particular structure of the model. In this case, we argued that if one can find a feedback controller which makes the closed-loop dynamics ISS (or some sort of ISS, i.e., semiglobal practical ISS, LiISS, etc.), one can complement the ISS controller with a model-free learning algorithm to learn the best estimates of the parametric uncertainties. Afterwards, to be able to propose more constructive proofs and results, we decided to focus on the well-known class of nonlinear models affine in the control vector. For this class of systems, we have proposed a constructive proof to design an ISS feedback controller, and then complemented it with a multivariable extremum-seeking algorithm, to learn online the values of the uncertainties. As we said earlier, we decided to call the learning approach "model-free" for the simple reason that it only requires the measurements of an output signal from the system. These measurements are compared to a desired reference signal (independent of the model) to learn the best estimates of the model uncertainties. We have guaranteed the stability of the proposed approach, simply by ensuring that the model-based robust controller leads to an ISS result which guarantees boundedness of the states of the closed-loop system, even during the learning phase. The ISS result together with a convergent learning algorithm eventually lead to a bounded output tracking error, which decreases with the decrease of the estimation error.

We believe that one of the main advantages of the proposed controller, compared to the existing model-based adaptive controllers, is the fact that we can learn (estimate) multiple uncertainties at the same time, even if they appear in the model equation in a challenging structure, for example, linearly dependent uncertainties affecting only one output, or uncertainties appearing in a nonlinear term of the model, which are well-known limitations of the model-based approaches. Another advantage of the proposed approach is that due to its modular design, one could easily change the learning algorithm without having to change the model-based part of the controller. Indeed, as long as the first part of the controller, that is, the model-based part, has been designed with a proper ISS property, one can plug into it any convergent learning model-free algorithm. For instance, one could use extremum-seeking algorithms with global or semiglobal convergence properties, rather than the local extremum-seeking algorithms used in this work. One could use reinforcement learning algorithms as well, and so on.

We ended this chapter by reporting the results obtained when applying the proposed approach to two different mechatronic examples. The first example is the electromagnetic actuator system, which is modeled by a general nonlinear equation. The second example is the multilink rigid robot, which is a well-known example of nonlinear models affine in the control.

REFERENCES

Adetola, V., Guay, M., 2007. Parameter convergence in adaptive extremum-seeking control. Automatica 43, 105–110.

Ariyur, K.B., Krstic, M., 2002. Multivariable extremum seeking feedback: analysis and design. In: Proceedings of the Mathematical Theory of Networks and Systems, South Bend, IN.

Ariyur, K.B., Krstic, M., 2003. Real-Time Optimization by Extremum-Seeking Control. John Wiley & Sons, Inc., New York, NY, USA.

Ariyur, K.B., Ganguli, S., Enns, D.F., 2009. Extremum seeking for model reference adaptive control. In: Proceedings of the AIAA Guidance, Navigation, and Control Conference, Chicago, IL.

Atinc, G., Benosman, M., 2013. Nonlinear backstepping learning-based adaptive control of electromagnetic actuators with proof of stability. In: IEEE 52nd Annual Conference on Decision and Control (CDC), 2013, pp. 1277–1282.

Benosman, M., 2013. Nonlinear learning-based adaptive control for electromagnetic actuators. In: IEEE European Control Conference, pp. 2904–2909.

Benosman, M., 2014a. Extremum-seeking based adaptive control for nonlinear systems. In: IFAC World Congress, Cape Town, South Africa, pp. 401–406.

Benosman, M., 2014b. Learning-based adaptive control for nonlinear systems. In: IEEE European Control Conference, Strasbourg, FR, pp. 920–925.

Benosman, M., Atinc, G., 2013a. Multi-parametric extremum seeking-based learning control for electromagnetic actuators. In: IEEE, American Control Conference, Washington, DC, pp. 1914–1919.

Benosman, M., Atinc, G., 2013b. Nonlinear adaptive control of electromagnetic actuators. In: SIAM Conference on Control and Applications, pp. 29–36.

Benosman, M., Atinc, G., 2013c. Nonlinear learning-based adaptive control for electromagnetic actuators. In: IEEE, European Control Conference, Zurich, pp. 2904–2909.

Benosman, M., Atinc, G., 2015a. Non-linear adaptive control for electromagnetic actuators. IET Control Theory Appl. 9 (2), 258–269.

Benosman, M., Atinc, G., 2015b. Nonlinear backstepping learning-based adaptive control of electromagnetic actuators. Int. J. Control 88 (3), 517–530.

Benosman, M., Xia, M., 2015. Extremum seeking-based indirect adaptive control for non-linear systems with time-varying uncertainties. In: IEEE, European Control Conference, Linz, Austria, pp. 2780–2785.

Benosman, M., Liao, F., Lum, K.Y., Wang, J.L., 2009. Nonlinear control allocation for non-minimum phase systems. IEEE Trans. Control Syst. Technol. 17 (2), 394–404.

Buffington, J.M., 1999. Modular control design for the innovative control effectors (ICE) tailless fighter aircraft configuration 101-3. Tech. Rep. AFRL-VA-WP-TR-1999-3057, U.S. Air Force Research Laboratory, Wright-Patterson AFB, OH.

Elmali, H., Olgac, N., 1992. Robust output tracking control of nonlinear MIMO systems via sliding mode technique. Automatica 28 (1), 145–151.

Guay, M., Zhang, T., 2003. Adaptive extremum seeking control of nonlinear dynamic systems with parametric uncertainties. Automatica 39, 1283–1293.

Guay, M., Dhaliwal, S., Dochain, D., 2013. A time-varying extremum-seeking control approach. In: IEEE, American Control Conference, pp. 2643–2648.

Haghi, P., Ariyur, K., 2011. On the extremum seeking of model reference adaptive control in higher-dimensional systems. In: IEEE, American Control Conference, pp. 1176–1181.

Hudon, N., Guay, M., Perrier, M., Dochain, D., 2008. Adaptive extremum-seeking control of convection-reaction distributed reactor with limited actuation. Comput. Chem. Eng. 32 (12), 2994–3001.

Ito, H., Jiang, Z., 2009. Necessary and sufficient small gain conditions for integral input-to-state stable systems: a Lyapunov perspective. IEEE Trans. Autom. Control 54 (10), 2389–2404.

Khalil, H., 2002. Nonlinear Systems, third ed. Prentice Hall, Englewood Cliffs, NJ.

Krstic, M., 2000. Performance improvement and limitations in extremum seeking. Syst. Control Lett. 39, 313–326.

Krstic, M., Wang, H.H., 2000. Stability of extremum seeking feedback for general nonlinear dynamic systems. Automatica 36, 595–601.

Krstic, M., Kanellakopoulos, I., Kokotovic, P., 1995. Nonlinear and Adaptive Control Design. John Wiley & Sons, New York.

Landau, I.D., Lozano, R., M'Saad, M., Karimi, A., 2011. Adaptive control: Algorithms, analysis and applications, Communications and Control Engineering. Springer-Verlag, Berlin.

Malisoff, M., Mazenc, F., 2005. Further remarks on strict input-to-state stable Lyapunov functions for time-varying systems. Automatica 41 (11), 1973–1978.

Moase, W., Manzie, C., Brear, M., 2009. Newton-like extremum seeking part I: theory. In: IEEE, Conference on Decision and Control, pp. 3839–3844.

Nesic, D., 2009. Extremum seeking control: convergence analysis. Eur. J. Control 15 (3-4), 331–347.

Nesic, D., Mohammadi, A., Manzie, C., 2013. A framework for extremum seeking control of systems with parameter uncertainties. IEEE Trans. Autom. Control 58 (2), 435–448.

Noase, W., Tan, Y., Nesic, D., Manzie, C., 2011. Non-local stability of a multi-variable extremum-seeking scheme. In: IEEE, Australian Control Conference, pp. 38–43.

Peterson, K., Stefanopoulou, A., 2004. Extremum seeking control for soft landing of electromechanical valve actuator. Automatica 40, 1063–1069.

Rotea, M., 2000. Analysis of multivariable extremum seeking algorithms. In: Proceedings of the American Control Conference, 2000, vol. 1, pp. 433–437.

Scheinker, A., 2013. Simultaneous stabilization of and optimization of unknown time-varying systems. In: IEEE, American Control Conference, pp. 2643–2648.

Spong, M.W., 1992. On the robust control of robot manipulators. IEEE Trans. Autom. Control 37 (11), 1782–2786.

Tan, Y., Nesic, D., Mareels, I., 2006. On non-local stability properties of extremum seeking control. Automatica 42, 889–903.

Teel, A.R., Popovic, D., 2001. Solving smooth and nonsmooth multivariable extremum seeking problems by the methods of nonlinear programming. In: Proceedings of the 2001 American Control Conference, 2001, vol. 3. IEEE, pp. 2394–2399.

Wang, C., Hill, D., 2006. Deterministic learning theory for identification, recognition, and control, Automation and Control Engineering Series. Taylor & Francis Group, Boca Raton.

Zhang, C., Ordóñez, R., 2012. Extremum-Seeking Control and Applications. Springer-Verlag, London.

Zhang, T., Guay, M., Dochain, D., 2003. Adaptive extremum seeking control of continuous stirred-tank bioreactors. AIChE J. 49, 113–123.

CHAPTER 5

Extremum Seeking-Based Real-Time Parametric Identification for Nonlinear Systems

5.1 INTRODUCTION

System identification can be interpreted as the science of learning the best possible model of a system given a set of experimental data. System identification has many ramifications. Indeed, it can be classified as system identification for linear versus nonlinear models, time domain based versus frequency domain, open–loop versus closed-loop, stochastic versus deterministic, identification for control versus identification for simulation and prediction, goal-oriented-based identification or codesign identification. This makes it difficult to write an exhaustive presentation of all the existing results in a few pages. Instead, we refer the reader to some outstanding surveys of the field, for example, Astrom and Eykhoff (1971), Ljung and Vicino (2005), Gevers (2006), Ljung (2010), and Pillonetto et al. (2014).

We will focus here on one small part of system identification, namely, system identification for nonlinear systems in state-space form. In this sub-area of system identification, we will present some results on deterministic approaches for both open-loop and closed-loop parametric identification in the time domain. This chapter does not pretend to be a detailed survey of nonlinear system identification, but rather a presentation of some new results in this area using model-free learning algorithms for parametric identification of nonlinear systems.

Indeed, identification for nonlinear systems is a challenging task which has been studied since the beginning by the research community in the context of nonlinear input/output mapping identification, a.k.a., black box identification, see for example, Sjoberg et al. (1995), Juditsky et al. (1995), Bendat (1990), Narendra and Parthasarathy (1990), Leontaritis and

Learning-Based Adaptive Control
http://dx.doi.org/10.1016/B978-0-12-803136-0.00005-1

Billings (1985), and Rangan et al. (1995). However, more recently the effort in nonlinear identification shifted toward the identification of parameters in nonlinear models dictated by the physics of the system, where the fundamental laws of physics are used to define the structure of the model, for example, nonlinear ordinary differential equations (ODEs) or nonlinear partial differential equations (PDEs), and then the identification techniques are used to find the best values of the coefficients for these models, see, Schittkowski (2002) and Klein and Morelli (2006). For example, state-space representations affine in the unknown coefficients have been used in Andrieu et al. (2004) and Schön et al. (2006), where an expectation maximization algorithm is used, together with a particle filtering approach, to identify the parameters. More results on using particle filtering to identify parameters in nonlinear models can be found in Kitagawa (1998), Schon and Gustafsson (2003), Doucet and Tadic (2003), and Andrieu et al. (2004). We also cite the series of papers by Glad and Ljung (1990a,b) and Ljung and Glad (1994), where differential algebraic tools are used for convexification of the identification problem for a class of nonlinear systems.

In this chapter, we follow the deterministic formulation of Ljung and Glad (1994), in the sense that we deal with nonlinear models in the deterministic time domain without disturbances. We simply formulate the identification problem as a minimization of a deterministic perfor-mance cost function, and use the extremum-seeking (ES) theory to solve the optimization problem online, leading to a simple real-time solution for open-loop and closed-loop parametric identification for nonlinear systems.

Another problem closely related to model identification, at least in our point of view, is the so-called stable model reduction for infinite-dimension systems. Indeed, the problem of reducing a PDE model to a simpler finite dimension ODE is of paramount importance in engineering and physics, where solving such PDE models is too time consuming. The idea of being able to reduce the PDE model to its simplest possible realization, without losing the main characteristics of the original model, such as stability and prediction precision, is appealing for any real-time model-based computations. However, this problem remains challenging because model reduction often comes at the cost of stability loss and prediction degradation. To remedy this problem, the research community, mainly in physics and applied mathematics, has been developing methods aiming at what is known as stable model reduction.

The results in this field are numerous, so we do not pretend here to fully cover the topic of stable model reduction; instead, we cite some references that we believe are relevant to our results in this field, which we will introduce later in this chapter. For instance in Cordier et al. (2013), a reduced order modeling method is proposed for stable model reduction of Navier-Stokes flow models. The main idea proposed by the authors is adding a nonlinear viscosity stabilizing term to the reduced order model (ROM). Then, the coefficients of the stabilizing nonlinear term are identified using a variational data-assimilation approach, which is based on solving a deterministic optimization problem. In Balajewicz et al. (2013) and Balajewicz (2013), a Lyapunov-based stable model reduction is proposed for incompressible flows. The approach is based on an iterative search of the projection modes satisfying some local Lyapunov stability condition. We also cite the references (San and Borggaard, 2014; San and Iliescu, 2013), where the case of stable model reduction for the Burgers' PDE is studied. The method proposed there is based on the so-called closure models, which boils down basically to the modification of some coefficients of the reduced order ODE model, with either constant additive terms (e.g., the constant eddy viscosity model) or time- and space-varying terms (e.g., Smagorinsky models). The added terms' amplitudes are tuned in such a way to stabilize the ROM.

In this chapter, we apply our ES-based identification approach to the problem of stable model reduction. Indeed, we see the problem of stable model reduction as a problem of coefficients' identification, either of the true coefficients of the model, or of the optimal values of the stabilizing closure models' coefficients. In this sense, the stable model reduction problem will be formulated as a problem of model parametric identification, and then will be solved in the context of an multiparametric extremum-seeking (MES)-based real-time identification approach.

This chapter is organized as follows: In Section 5.2 some general notations and tools will be recalled for the completeness of this chapter. In Section 5.3, we present the ES-based open-loop identification for nonlinear systems. Next, in Section 5.4, we study the problem of closed-loop parametric identification for nonlinear systems. Section 5.5 is dedicated to the challenging problem of PDE models' stable reduction and identification. Numerical examples on two mechatronics systems and one fluid mechanics problem are reported in Section 5.6. Finally, conclusions and open research problems are discussed in Section 5.7.

5.2 BASIC NOTATIONS AND DEFINITIONS

Throughout the chapter we will use $\| \cdot \|$ to denote the Euclidean vector norm; that is, for $x \in \mathbb{R}^n$ we have $\|x\| = \sqrt{x^T x}$. The Kronecker delta function is defined as $\delta_{ij} = 0$, for $i \neq j$ and $\delta_{ii} = 1$. The Frobenius norm of a matrix $A \in \mathbb{R}^{m \times n}$, with elements a_{ij}, is defined as $\|A\|_F \triangleq \sqrt{\sum_{i=1}^{n} \sum_{j=1}^{n} |a_{ij}|^2}$. Similarly, the Frobenius norm of a tensor $A \in \mathbb{R}^{\otimes_i n_i}$, with elements $a_{\mathbf{i}} = a_{i_1 \cdots i_k}$, is defined as $\|A\|_F \triangleq \sqrt{\sum_{\mathbf{i}=1}^{\mathbf{n}} |a_{\mathbf{i}}|^2}$. The 1-norm of $x \in \mathbb{R}^n$ is denoted by $\|x\|_1$. We use the following norm properties (e.g., Golub and Van Loan, 1996):

1. for any $x \in \mathbb{R}^n$, $\|x\| \leq \|x\|_1$;
2. for any $x \in \mathbb{R}^n$, $A \in \mathbb{R}^{m \times n}$, $\|Ax\| \leq \|A\|_F \|x\|$;
3. for any $x, y \in \mathbb{R}^n$, $\|x\| - \|y\| \leq \|x - y\|$; and
4. for any $x, y \in \mathbb{R}^n$, $x^T y \leq \|x\| \|y\|$.

For an $n \times n$ matrix P, we write $P > 0$ if it is positive definite. Similarly, we write $P < 0$ if it is negative definite. We use $\text{diag}\{A_1, A_2, \ldots, A_n\}$ to denote a diagonal block matrix with n blocks. For a matrix B, we denote $B(i, j)$ as the (i, j) element of B. We denote by I_n the identity matrix or simply I if the dimension is clear from the context; we denote $\mathbf{0}$ the vector $[0, \ldots, 0]^T$ of appropriate dimension. We denote by $\text{Max}(v), v \in \mathbb{R}^n$ the maximum element of the vector v. We will use $(\dot{\cdot})$ for the short notation of time derivative, and x^T for the transpose of a vector x. We denote by $\mathcal{C}^k$ functions that are k times differentiable. A function is said to be analytic in a given set if it admits a convergent Taylor series approximation in some neighborhood of every point of the set. We denote by $f_1 \circ f_2$ the composition of the two mappings f_1 and f_2. A continuous function $\alpha \colon [0, a) \to [0, \infty)$ is said to belong to class $\mathcal{K}$ if it is strictly increasing and $\alpha(0) = 0$. It is said to belong to class $\mathcal{K}_\infty$ if $a = \infty$ and $\alpha(r) \to \infty$ as $r \to \infty$. A continuous function $\beta \colon [0, a) \times [0, \infty) \to [0, \infty)$ is said to belong to class $\mathcal{KL}$ if, for each fixed s, the mapping $\beta(r, s)$ belongs to class $\mathcal{K}$ with respect to r and, for each fixed r, the mapping $\beta(r, s)$ is decreasing with respect to s and $\beta(r, s) \to 0$ as $s \to \infty$. We consider the Hilbert space $\mathcal{Z} = L^2([0, 1])$, which is the space of Lebesgue square integrable functions, that is, $f \in \mathcal{Z}$, if $\int_0^1 |f(x)|^2 dx < \infty$. We define on $\mathcal{Z}$ the inner product $\langle \cdot, \cdot \rangle_{\mathcal{Z}}$ and the associated norm $\| \cdot \|_{\mathcal{Z}}$, as $\|f\|_{\mathcal{Z}}^2 = \int_0^1 |f(x)|^2 dx$, and $\langle f, g \rangle_{\mathcal{Z}} = \int_0^1 f(x)g(x)dx$, for $f, g \in \mathcal{Z}$. A function $\omega(t, x)$ is in $L^2([0, T]; \mathcal{Z})$ if for each $0 \leq t \leq T$, $\omega(t, \cdot) \in \mathcal{Z}$, and $\int_0^T \|\omega(t, \cdot)\|_{\mathcal{Z}}^2 dt \leq \infty$.

Let us, for completeness of this chapter, recall some stability definitions for nonlinear systems (cf. Chapter 1). Consider the general time-varying system

$$\dot{x} = f(t, x, u), \tag{5.1}$$

where $f \colon [0, \infty) \times \mathbb{R}^n \times \mathbb{R}^{n_a} \to \mathbb{R}^n$ is piecewise continuous in t and locally Lipschitz in x and u, uniformly in t. The input $u(t)$ is a piecewise continuous, bounded function of t for all $t \geq 0$.

Definition 5.1 (Khalil, 2002). *The system* (5.1) *is said to be input-to-state stable (ISS) if there exist a class $\mathcal{KL}$ function β and a class $\mathcal{K}$ function γ such that for any initial state $x(t_0)$ and any bounded input $u(t)$, the solution $x(t)$ exists for all $t \geq t_0$ and satisfies*

$$\|x(t)\| \leq \beta(\|x(t_0)\|, t - t_0) + \gamma\left(\sup_{t_0 \leq \tau \leq t} \|u(\tau)\|\right).$$

Definition 5.2 (Khalil, 2002; Sontag, 2008). *The system* (5.1) *associated with the output mapping $y = h(t, x, u)$, where $h \colon [0, \infty) \times \mathbb{R}^n \times \mathbb{R}^{n_a} \to \mathbb{R}^n$ is piecewise continuous in t and continuous in x and u, is said to be input-to-output stable (IOS) if there exist a class $\mathcal{KL}$ function β and a class $\mathcal{K}$ function γ such that for any initial state $x(t_0)$ and any bounded input $u(t)$, the solution $x(t)$ exists for all $t \geq t_0$; that is, y exists for all $t \geq t_0$ and satisfies*

$$\|y(t)\| \leq \beta(\|x(t_0)\|, t - t_0) + \gamma\left(\sup_{t_0 \leq \tau \leq t} \|u(\tau)\|\right).$$

Theorem 5.1 (Khalil, 2002; Malisoff and Mazenc, 2005). *Let $V \colon [0, \infty) \times \mathbb{R}^n \to \mathbb{R}$ be a continuously differentiable function such that*

$$\alpha_1(\|x\|) \leq V(t, x) \leq \alpha_2(\|x\|)$$

$$\frac{\partial V}{\partial t} + \frac{\partial V}{\partial x} f(t, x, u) \leq -W(x), \quad \forall \|x\| \geq \rho(\|u\|) > 0 \qquad (5.2)$$

for all $(t, x, u) \in [0, \infty) \times \mathbb{R}^n \times \mathbb{R}^m$, where α_1, α_2 are class $\mathcal{K}_\infty$ functions, ρ is a class $\mathcal{K}$ function, and $W(x)$ is a continuous positive definite function on $\mathbb{R}^n$. Then, the system (5.1) *is ISS.*

Definition 5.3 (Haddad and Chellaboina, 2008). *A system $\dot{x} = f(t, x)$ is said to be Lagrange stable if for every initial condition x_0 associated with the time instant t_0, there exists $\epsilon(x_0)$, such that $\|x(t)\| < \epsilon, \forall t \geq t_0 \geq 0$.*

5.3 ES-BASED OPEN-LOOP PARAMETRIC IDENTIFICATION FOR NONLINEAR SYSTEMS

5.3.1 Problem Formulation

In this section we consider general nonlinear systems of the form

$$\begin{aligned} \dot{x} &= f(t, x, u, \theta), \quad x(0) = x_0, \\ y &= h(x, u, \theta), \end{aligned} \qquad (5.3)$$

where $x \in \mathbb{R}^n$, $u \in \mathbb{R}^{n_a}$, and $y \in \mathbb{R}^m (n_a \geq m)$, represent the state, the control input, and the observed output vectors, respectively; x_0 is a given finite initial condition; $\theta \in \mathbb{R}^{n_\theta}$ represents an n_θ-dimensional vector of parameters to be identified. The vector fields f and the function h satisfy the following assumptions.

Assumption 5.1. $f \colon \mathbb{R} \times \mathbb{R}^n \times \mathbb{R}^{n_a} \times \mathbb{R}^{n_\theta} \to \mathbb{R}^n$ *is a* $\mathcal{C}^\infty$ *vector field on a bounded set* X *of* $\mathbb{R}^n \times \mathbb{R}^{n_a} \times \mathbb{R}^{n_\theta}$ *and* h *is a* $\mathcal{C}^\infty$ *function on* X.

Assumption 5.2. *We assume the case where* θ *is a vector of some physical parameters of the system, and that its nominal (not exact) value* θ_0 *is known.*

Remark 5.1. *Assumption 5.2 makes sense in real applications, where the engineers often have some knowledge of nominal values for the physical parameters of the system that they are trying to model. However, these nominal values are often inexact and need to be fine-tuned; that is, their exact values have to be estimated.*

Remark 5.2. *Some readers might find the problem formulation given by Eq. (5.3) "lacking of noise." Indeed, we have chosen a classical deterministic formulation, frequently used by other researchers in the field of deterministic identification approaches, see for example, Ljung and Glad (1994). Moreover, the learning-based identification approach which we are proposing here is intrinsically robust to bounded measurement noises. We will come back to this point later in this chapter.*

5.3.2 Open-Loop Parameters Estimation

To proceed with an open-loop identification approach, we need the following assumption on the system.

Assumption 5.3. *We assume that the system (5.3) is Lagrange stable.*

Remark 5.3. *From the Lagrange stability assumption and the smoothness assumptions on the model, it is straightforward to conclude about the boundedness of the output function* h, *for any bounded input vector* u *and bounded parameter vector* θ.

Remark 5.4. *The Lagrange stability assumption is clearly needed for open-loop identification, to be able to run the system in open loop and compare its observable bounded output to its model-based output estimate.*

Let us define the output estimation error

$$e_y(t) = y(t) - y_m(t), \tag{5.4}$$

where $y_m(t)$ represents the output measurement at time instant t.

Let us define now the following identification cost function:

$$Q(\hat{\theta}) = F(e_y(\hat{\theta})), \tag{5.5}$$

where $\hat{\theta}$ is the estimate of the parameter vector θ, and $F\colon \mathbb{R}^m \rightarrow \mathbb{R}$, $F(0) = 0, F(e_y) > 0$, for $e_y \neq 0$.

This cost function will be used as a learning cost function. To do so, we need to state the following assumption on Q.

Assumption 5.4. *The cost function Q has a local minimum at $\hat{\theta}^* = \theta$.*

Assumption 5.5. *The original parameters' estimate vector $\hat{\theta}(0)$, that is, the nominal parameters' value, is close enough to the actual parameters' vector θ.*

Assumption 5.6. *The cost function is analytic and its variation with respect to the uncertain variables is bounded in the neighborhood of θ^*, that is, $\|\frac{\partial Q}{\partial \theta}(\tilde{\theta})\| \leq \xi_2, \xi_2 > 0, \tilde{\theta} \in \mathcal{V}(\theta^*)$, where $\mathcal{V}(\theta^*)$ denotes a compact neighborhood of θ^*.*

Based on this open-loop formulation, we can now propose a simple MES-based identification algorithm.

Theorem 5.2. *Consider the system (5.3), satisfying Assumption 5.3, with bounded open-loop control $u(t)$. Then, the parameter vector θ can be estimated starting from its nominal value θ_{nom}, such that*

$$\hat{\theta}(t) = \theta_{nom} + \Delta\theta(t), \tag{5.6}$$

where $\Delta\theta = [\delta\theta_1, \ldots, \delta\theta_{n_\theta}]^T$ is computed using the MES algorithm

$$\dot{z}_i = a_i \sin\left(\omega_i t + \frac{\pi}{2}\right) Q(\hat{\theta})$$

$$\delta\theta_i = z_i + a_i \sin\left(\omega_i t - \frac{\pi}{2}\right), \quad i \in \{1, \ldots, n_\theta\} \tag{5.7}$$

with $\omega_i \neq \omega_j, \omega_i + \omega_j \neq \omega_k, i, j, k \in \{1, \ldots, n_\theta\}$, and $\omega_i > \omega^*, \forall i \in \{1, \ldots, n_\theta\}$, with ω^* large enough, and Q given by Eq. (5.5). Furthermore, under Assumptions 5.4–5.6, the norm of the estimation error $e_\theta = \theta - \hat{\theta}(t)$ admits the following bound:

$$\|e_\theta(t)\| \leq \frac{\xi_1}{\omega_0} + \sqrt{\sum_{i=1}^{i=n_\theta} a_i^2}, t \rightarrow \infty \tag{5.8}$$

where $\xi_1 > 0$ and $\omega_0 = \max_{i \in \{1, \ldots, n_\theta\}} \omega_i$.

Proof. First, based on Assumptions 5.4–5.6, the ES nonlinear dynamics (5.54) can be approximated by a linear averaged dynamic (using averaging approximation over time, Rotea, 2000a, p. 435, Definition 1). Furthermore, $\exists \xi_1, \omega^*$, such that for all $\omega_0 = \max_{i \in \{1, \ldots, n_\theta\}} \omega_i > \omega^*$, the solution of the averaged model $\Delta\theta_{aver}(t)$ is locally close to the solution of the original MES dynamics, and satisfies (Rotea, 2000a, p. 436)

$$\|\Delta\theta(t) - d(t) - \Delta\theta_{aver}(t)\| \leq \frac{\xi_1}{\omega_0}, \xi_1 > 0, \quad \forall t \geq 0$$

with $d(t) = (a_1 \sin(\omega_1 t - \frac{\pi}{2}), \ldots, a_{n_\theta} \sin(\omega_{n_\theta} t - \frac{\pi}{2}))^T$. Moreover, because Q is analytic it can be approximated locally in $\mathcal{V}(\beta^*)$ with a Taylor series up to second order, which leads to (Rotea, 2000a, p. 437)

$$\lim_{t \to \infty} \Delta\theta_{\text{aver}}(t) = \Delta\theta^*,$$

such that

$$\Delta\theta^* + \theta_{\text{nom}} = \theta,$$

which together with the previous inequality leads to

$$\|\Delta\theta(t) - \Delta\theta^*\| - \|d(t)\| \le \|\Delta\theta(t) - \Delta\theta^* - d(t)\|$$

$$\le \frac{\xi_1}{\omega_0}, \quad \xi_1 > 0, t \to \infty$$

$$\Rightarrow \|\Delta\theta(t) - \Delta\theta^*\| \le \frac{\xi_1}{\omega_0} + \|d(t)\|, \quad t \to \infty.$$

This finally implies that

$$\|\Delta\theta(t) - \Delta\theta^*\| \le \frac{\xi_1}{\omega_0} + \sqrt{\sum_{i=1}^{i=n_\theta} a_i^2}, \quad \xi_1 > 0, \qquad t \to \infty.$$

The previous theorem is written in continuous time domain, where the excitation signal u is assumed to be bounded over an infinite time support, that is, $t \to \infty$. However, in real applications we want to excite the system with a bounded signal over a finite time support, that is, $t \in [t_0, t_f], 0 \le t_0 < t_f$. In this case, the identification convergence results can be formulated in the following lemma.

Lemma 5.1. *Consider the system* (5.3), *satisfying Assumption 5.3, under bounded open-loop control* $u(t)$, *over a finite time support, that is,* $t \in [t_0, t_f], 0 \le t_0 < t_f$. *Then, the parameter vector* θ *can be iteratively estimated starting from its nominal value* θ_{nom}, *such that*

$$\hat{\theta}(t) = \theta_{\text{nom}} + \Delta\theta(t),$$
$$\Delta\theta(t) = \hat{\Delta\theta}((I-1)t_f), \quad (I-1)t_f \le t < It_f, \qquad I = 1, 2, \ldots \tag{5.9}$$

where I *is the iteration number and* $\hat{\Delta\theta} = [\hat{\delta\theta}_1, \ldots, \hat{\delta\theta}_{n_\theta}]^T$ *is computed using the iterative MES algorithm*

$$\dot{z}_i = a_i \sin\left(\omega_i t + \frac{\pi}{2}\right) Q(\hat{\theta})$$
$$\delta\theta_i = z_i + a_i \sin\left(\omega_i t - \frac{\pi}{2}\right), \quad i \in \{1, \ldots, n_\theta\} \tag{5.10}$$

with Q given by Eq. (5.5), $\omega_i \neq \omega_j, \omega_i + \omega_j \neq \omega_k, i, j, k \in \{1, \ldots, n_\theta\}$, and $\omega_i > \omega^*, \forall i \in \{1, \ldots, n_\theta\}$, with ω^* large enough. Then under Assumptions 5.12–5.6, the estimate upper bound is given by

$$\|\hat{\theta}(It_f) - \theta\| \leq \frac{\xi_1}{\omega_0} + \sqrt{\sum_{i=1}^{i=n_\theta} a_i^2}, \quad I \to \infty$$

where $\xi_1 > 0$ and $\omega_0 = \max_{i \in \{1, \ldots, n_\theta\}} \omega_i$.

Remark 5.5. *As we mentioned earlier, we chose here to write the identification problem in a deterministic formulation, without explicitly adding to the system measurement noises. Indeed, the reason behind this is simply that the perturbation-based MES algorithms are well known to be robust to measurement noise; see for example, the comparative study in robotics reported in Calli et al. (2012). Moreover, more "sophisticated" MES algorithms could easily be substituted instead of the first-order MES algorithm (5.10) in cases where the measurement noise level is high. For instance, in Krstic (2000), several filters have been added to the perturbation-based ES algorithm to reject additive measurement noises. In Stankovica and Stipanovicb (2010), the case of stochastic measurement noises has been studied in relation to perturbation-based ES algorithms.*

Remark 5.6. *Assumptions 5.4 and 5.5 mean that the search for the parameters' actual values is local. This type of assumption is well accepted in the identification community, and corresponds to the so-called "local identifiability" assumption, see for example, Gevers (2006). Indeed, in real applications it is often the case where the engineers have some initial idea about the values of the system's physical parameters, for example, mass, inertia, and so on. In this case, the identification process boils down to fine-tuning these parameters in the neighborhood of the nominal values, which corresponds to local identification.*

Remark 5.7. *It is well known in classical identification algorithms that the regressor matrix has to satisfy a persistent excitation (PE) condition to ensure the convergence of the identified parameters to the actual system's parameters. It might seem that the MES-based algorithm does not require any PE condition. However, if we examine Eq. (5.10) we can see that in our case the identification output is the cost function Q, and thus the regressor matrix in this case will be the diagonal matrix whose elements are the perturbation signals $a_i \sin(\omega_i t + \frac{\pi}{2})$, which clearly satisfies a PE condition (cf. Chapter 1).*

5.4 ES-BASED CLOSED-LOOP PARAMETRIC IDENTIFICATION FOR NONLINEAR SYSTEMS

In the previous section we considered the case of open-loop identification, which requires the system to be stable in open loop. Unfortunately, many real-life systems do not exhibit such a nice open-loop behavior. In such cases, it is necessary to close the loop on the system to stabilize the system first, before attempting to identify its parameters. In this section we will present some algorithms which deal with this problem. Another advantage of closed-loop identification versus open-loop identification is its robustness with respect to initial condition errors. Indeed, open-loop identification algorithms are sensitive to initial condition mismatches because there is no way to guarantee any convergence of states' trajectories starting from different initial conditions. On the contrary, using a feedback controller to stabilize the system around a given equilibrium point, or equivalently around a desired trajectory, ensures, at least locally, robustness with respect to initial condition errors.

5.4.1 Problem Formulation

Let us consider again the system in Eq. (5.3), but where the stability Assumption 5.3 does not hold anymore. Instead, we pose the problem as an output tracking problem, and ensure some IOS property by a proper design of a feedback control term. In this context, the new input to the identification problem is the desired output trajectory.

To formalize this idea, we introduce a rather general assumption.

Let us consider a bounded smooth output desired time trajectory $y_{\text{ref}}: [0, \infty) \to \mathbb{R}^m$. Then, we assume the following.

Assumption 5.7. *There exists a control feedback $u_{iss}(t, x, y_{ref}, \hat{\theta}): \mathbb{R} \times \mathbb{R}^n \times \mathbb{R}^m \times \mathbb{R}^{n_\theta} \to \mathbb{R}^m$, with $\hat{\theta}$ being the estimate of the vector parameter θ, such that the closed-loop dynamics associated with Eq. (5.3)*

$$\dot{x} = f \circ u_{iss} \equiv F(t, x, y_{ref}, e_\theta), \quad e_\theta = \theta - \hat{\theta},$$
$$y_{cl} = h \circ u_{iss} \equiv H(t, x, y_{ref}, e_\theta), \tag{5.11}$$

is IOS from the input vector (y_{ref}^T, e_θ^T) to the new output vector y_{cl}.

Remark 5.8. *We explicitly wrote the feedback control as a function of the vector y_{ref}, even though it could have been included (implicitly) in the time-varying term. The reason for this "abuse of notation" is that we want to emphasize the fact that in this formulation, the input to the "new" stable system to be identified is y_{ref}.*

Remark 5.9. *Assumption 5.7 imposes boundedness of the output vector y_{cl}, for bounded input vector (y_{ref}^T, e_θ^T). The boundedness of the state vector of the closed-loop dynamics could be deduced from the boundedness of y_{cl} if the mapping H is radially unbounded with respect to the state vector, for example, class $\mathcal{K}_\infty$ mapping.*

Now that we have reformulated the problem as a parametric identification of the IOS system (5.11), we can state a similar estimation result to the previous open-loop case.

Theorem 5.3. *Consider the system (5.3), satisfying Assumption 5.7, under bounded open-loop input $y_{ref}(t)$. Then, the parameter vector θ can be estimated starting from its nominal value θ_{nom}, such that*

$$\hat{\theta}(t) = \theta_{nom} + \Delta\theta(t), \qquad (5.12)$$

where $\Delta\theta = [\delta\theta_1, \ldots, \delta\theta_{n_\theta}]^T$ is computed using the MES algorithm

$$\dot{z}_i = a_i \sin\left(\omega_i t + \frac{\pi}{2}\right) Q(\hat{\theta}),$$

$$\delta\theta_i = z_i + a_i \sin\left(\omega_i t - \frac{\pi}{2}\right), \quad i \in \{1, \ldots, n_\theta\} \qquad (5.13)$$

with $\omega_i \neq \omega_j, \omega_i + \omega_j \neq \omega_k, i, j, k \in \{1, \ldots, n_\theta\}$, and $\omega_i > \omega^*, \forall i \in \{1, \ldots, n_\theta\}$, with ω^* large enough and Q given by Eq. (5.5), with $e_y = y_d - y_{ref}$. Furthermore, under Assumptions 5.4–5.6, the norm of the estimation error $e_\theta = \theta - \hat{\theta}(t)$ admits the following bound:

$$\|e_\theta(t)\| \leq \frac{\xi_1}{\omega_0} + \sqrt{\sum_{i=1}^{i=n_\theta} a_i^2}, \quad t \to \infty \qquad (5.14)$$

where $\xi_1 > 0$ and $\omega_0 = \max_{i \in \{1, \ldots, n_\theta\}} \omega_i$.

Similar to the open-loop case, it is straightforward to write a convergence result for the practical case of trajectories y_{ref} with bounded time support.

Lemma 5.2. *Consider the system (5.3), satisfying Assumption 5.7, under bounded open-loop input $y_{ref}(t)$, over a finite time support, that is, $t \in [t_0, t_f], 0 \leq t_0 < t_f$. Then, the parameter vector θ can be iteratively estimated starting from its nominal value θ_{nom}, such that*

$$\hat{\theta}(t) = \theta_{nom} + \Delta\theta(t),$$

$$\Delta\theta(t) = \hat{\Delta\theta}((I-1)t_f), \quad (I-1)t_f \leq t < It_f, \qquad I = 1, 2, \ldots \qquad (5.15)$$

where I is the iteration number and $\hat{\Delta\theta} = [\hat{\delta\theta}_1, \ldots, \hat{\delta\theta}_{n_\theta}]^T$ is computed using the iterative MES algorithm

$$\dot{z}_i = a_i \sin\left(\omega_i t + \frac{\pi}{2}\right) Q(\hat{\theta}),$$

$$\hat{\delta\theta}_i = z_i + a_i \sin\left(\omega_i t - \frac{\pi}{2}\right), \quad i \in \{1, \ldots, n_\theta\} \qquad (5.16)$$

with Q given by Eq. (5.5) with $e_y = y_d - y_{ref}$, $\omega_i \neq \omega_j$, $\omega_i + \omega_j \neq \omega_k$, $i, j, k \in \{1, \ldots, n_\theta\}$, and $\omega_i > \omega^*$, $\forall i \in \{1, \ldots, n_\theta\}$, with ω^* large enough. Then, under Assumptions 5.4–5.6, the estimate upper bound writes as

$$\|\hat{\theta}(It_f) - \theta\| \leq \frac{\xi_1}{\omega_0} + \sqrt{\sum_{i=1}^{i=n_\theta} a_i^2}, \quad I \to \infty \qquad (5.17)$$

where $\xi_1 > 0$ and $\omega_0 = \max_{i \in \{1, \ldots, n_\theta\}} \omega_i$.

It is clear from the previous results that the same tools can be used to estimate the model parameters in the case of unstable open-loop dynamics, provided that Assumption 5.7 can be satisfied. In other words, the difficulty here is more about finding an IOS stabilizing feedback controller than identifying the parameters, which can then easily be done using the model-free MES algorithm (5.12), (5.13).

Although we cannot propose a controller which will render any nonlinear system IOS in the general case; however, it is possible to propose such feedback controllers in some specific class of nonlinear systems. We shall do that in the remainder of this section.

5.4.2 Parametric Estimation in the Case of Nonlinear Systems Affine in the Control

Now, we focus on a particular class of nonlinear systems (5.3), namely, the class of systems of the form

$$\dot{x} = f(x) + \Delta f(t, x) + g(x)u,$$
$$y = h(x), \qquad (5.18)$$

where $x \in \mathbb{R}^n$, $u \in \mathbb{R}^{n_a}$, and $y \in \mathbb{R}^m$ $(n_a \geq m)$ represent the state, the input, and the controlled output vectors, respectively. $\Delta f(t, x)$ is a vector field representing additive model uncertainties, that is, the part of the model with parametric uncertainties. The vector fields f, Δf, columns of g, and function h satisfy the following classical assumptions.

Assumption 5.8. *The function $f : \mathbb{R}^n \to \mathbb{R}^n$ and the columns of $g : \mathbb{R}^n \to \mathbb{R}^{n_a}$ are C^∞ vector fields on a bounded set X of $\mathbb{R}^n$ and $h : \mathbb{R}^n \to \mathbb{R}^m$ is a smooth class $\mathcal{K}_\infty$ function. The vector field $\Delta f(x)$ is C^1 on X.*

Assumption 5.9. *System (5.18) has a well-defined (vector) relative degree $\{r_1, \ldots, r_m\}$ at each point $x^0 \in X$, and the system is linearizable, that is, $\sum_{i=1}^m r_i = n$.*

The goal here is to construct a state feedback which makes the closed-loop dynamics bounded from the output reference signal to the closed-loop system's states.

To do so, we need one more assumption on the form of the uncertainty term Δf in Eq. (5.18).

Assumption 5.10. *The additive uncertainty term $\Delta f(t, x)$ in Eq. (5.18) appears as an additive uncertainty term in the input-output linearized model obtained as follows:*

$$y^{(r)}(t) = b(\xi(t)) + A(\xi(t))u(t) + \Delta b(t, \xi(t)), \qquad (5.19)$$

where $\Delta b(t, \xi)$ is C^1 with respect to the state vector $\xi \in \tilde{X}$, where

$$y^{(r)}(t) = [y_1^{(r_1)}(t), \ y_2^{(r_2)}(t), \dots, y_m^{(r_m)}(t)]^T,$$

$$\xi(t) = [\xi^1(t), \dots, \xi^m(t)]^T, \qquad (5.20)$$

$$\xi^i(t) = [y_i(t), \dots, y_i^{(r_i-1)}(t)], \quad 1 \le i \le m,$$

and, the functions $b(\xi)$, $A(\xi)$ can be written as functions of f, g, and h, and $A(\xi)$ is nonsingular in $\tilde{X}$, where $\tilde{X}$ is the image of the set of X by the diffeomorphism $x \to \xi$ between the states of system (5.18) and the linearized model (5.19), (5.20).

Based on Assumption 5.10, we will construct an IOS controller for the system (5.18), following a two-step procedure. First, we assume that there is no uncertain term in the model, and then write a nominal feedback controller for the nominal model. Second, we introduce back the uncertain term in the model and use Lyapunov reconstruction techniques to ensure IOS stability despite the uncertainty.

First, let us assume that $\Delta f \equiv 0$ in Eq. (5.18), that is, $\Delta b \equiv 0$ in Eq. (5.19). In this case the linearized model can be rewritten as

$$v(t) = b(\xi(t)) + A(\xi(t))u(t), \qquad (5.21)$$

$$y^{(r)}(t) = v(t). \qquad (5.22)$$

From Eqs. (5.21), (5.22) we write the following nominal output tracking controller:

$$u_\mathrm{n} = A^{-1}(\xi)\left[v_s(t, \xi) - b(\xi)\right], \qquad (5.23)$$

where v_s is an $m \times 1$ vector and the ith ($1 \le i \le m$) element v_{si} is given by

$$v_{si} = y_{\mathrm{ref}_i}^{(r_i)} - K_{r_i}^i \left(y_i^{(r_i-1)} - y_{\mathrm{ref}_i}^{(r_i-1)}\right) - \cdots - K_1^i(y_i - y_{\mathrm{ref}_i}). \qquad (5.24)$$

We finally obtain the following tracking dynamics:

$$e_i^{(r_i)}(t) + K_{r_i}^i e^{(r_i-1)}(t) + \cdots + K_1^i e_i(t) = 0, \qquad (5.25)$$

where $e_i(t) = y_i(t) - y_{\mathrm{ref}_i}(t), i \in \{1, 2, \ldots, m\}$. To ensure the stability of the error dynamics, that is, of the nominal controller acting on the nominal model, we introduce the following straightforward assumption.

Assumption 5.11. *There exists a nonempty set $\mathcal{A}$ of gains $K_j^i \in \mathcal{A}, i \in \{1, 2, \ldots, m\}, j \in \{1, 2, \ldots, r_i\}$, such that the polynomials in Eq. (5.25) are Hurwitz.*

To use more familiar matrix notations we define the following vectors $z = [z^1, z^2, \ldots, z^m]^T$, where $z^i = [e_i, \dot{e}_i, \ldots, e_i^{(r_i-1)}]$ and $i \in \{1, 2, \ldots, m\}$, which leads to another form of Eq. (5.25)

$$\dot{z} = \tilde{A}z,$$

where $\tilde{A} \in \mathbb{R}^{n \times n}$ is a diagonal block matrix given by

$$\tilde{A} = \mathrm{diag}\{\tilde{A}_1, \tilde{A}_2, \ldots, \tilde{A}_m\}, \qquad (5.26)$$

and $\tilde{A}_i$, $1 \leq i \leq m$ are $r_i \times r_i$ matrices given by

$$\tilde{A}_i = \begin{bmatrix} 0 & 1 & & & \\ 0 & & 1 & & \\ 0 & & & \ddots & \\ \vdots & & & & 1 \\ -K_1^i & -K_2^i & \cdots & \cdots & -K_{r_i}^i \end{bmatrix}.$$

Based on Assumption 5.11, there exists a positive definite matrix $P > 0$ such that (see, e.g., Khalil, 2002)

$$\tilde{A}^T P + P\tilde{A} = -I. \qquad (5.27)$$

Next, we introduce back the uncertain term in the model, and build upon the nominal controller to write a controller ensuring IOS of the closed-loop uncertain system. To do so, we write the total controller as

$$u_f = u_n + u_r, \qquad (5.28)$$

where the nominal controller u_n is given by Eq. (5.23) and the robust controller u_r will be obtained based on Lyapunov reconstruction techniques. Indeed, by using Eqs. (5.19), (5.28) we can write

$$y^{(r)}(t) = b(\xi(t)) + A(\xi(t))u_f + \Delta b(t, \xi(t)),$$
$$= b(\xi(t)) + A(\xi(t))u_n + A(\xi(t))u_r + \Delta b(t, \xi(t)),$$
$$= v_s(t, \xi) + A(\xi(t))u_r + \Delta b(t, \xi(t)), \tag{5.29}$$

where Eq. (5.29) holds from Eq. (5.23). This leads to the following error dynamics:

$$\dot{z} = \tilde{A}z + \tilde{B}\delta, \tag{5.30}$$

where $\tilde{A}$ is defined in Eq. (5.26), δ is an $m \times 1$ vector given by

$$\delta = A(\xi(t))u_r + \Delta b(t, \xi(t)), \tag{5.31}$$

and the matrix $\tilde{B} \in \mathbb{R}^{n \times m}$ is given by

$$\tilde{B} = \begin{bmatrix} \tilde{B}_1 \\ \tilde{B}_2 \\ \vdots \\ \tilde{B}_m \end{bmatrix}, \tag{5.32}$$

where each $\tilde{B}_i$ $(1 \le i \le m)$ is given by an $r_i \times m$ matrix such that

$$\tilde{B}_i(l, q) = \begin{cases} 1 & \text{for } l = r_i, \ q = i, \\ 0 & \text{otherwise.} \end{cases}$$

If we choose $V(z) = z^T P z$ as a Lyapunov function for the dynamics (5.30), where P is the solution of the Lyapunov equation (5.27), we obtain

$$\dot{V}(t) = \frac{\partial V}{\partial z}\dot{z},$$
$$= z^T(\tilde{A}^T P + P\tilde{A})z + 2z^T P\tilde{B}\delta,$$
$$= -\|z\|^2 + 2z^T P\tilde{B}\delta, \tag{5.33}$$

where δ is given by Eq. (5.31).

Next, we study different cases based on the expression of the uncertain term Δb.

5.4.2.1 Case 1

We consider first the case where the uncertain term has the following form:

$$\Delta b(t, \xi(t)) = EL(\xi, t), \tag{5.34}$$

where $E \in \mathbb{R}^{m \times m}$ is a matrix of unknown constant parameters and $L(\xi, t) \colon \mathbb{R}^n \times \mathbb{R} \to \mathbb{R}^m$ is a known bounded function of states and time variables. Let us define the estimation error vector by $e_E = E - \hat{E}$, where $\hat{E}(t)$ denotes the estimate of E. To use vectorial notations, we define the unknown parameter vector $\theta = [E(1,1), \ldots, E(m,m)]^T \in \mathbb{R}^{m^2}$, that is, concatenation of all elements of E, and its estimate vector is denoted by $\hat{\theta}(t) = [\hat{E}(1,1), \ldots, \hat{E}(m,m)]^T$, which leads to the estimation error vector $e_\theta(t) = \theta - \hat{\theta}(t)$. To ensure that Assumption 5.7 holds, we propose the following feedback controller:

$$u_{\mathrm{r}} = -A^{-1}(\xi)[\tilde{B}^T P z \| L(\xi, t) \|^2 + \hat{E}(t) L(\xi, t)]. \qquad (5.35)$$

Now, we need to analyze the boundedness of the closed-loop dynamics obtained by applying the controller (5.23), (5.28), (5.35) to the system (5.19). To do so, we define the new output for the closed-loop system as $y_{\mathrm{cl}} = z$. Loosely speaking, we want to show that this output (and by inference under the present assumptions, the states x) remains bounded for any bounded input signals y_{ref} and e_θ. An easy way to prove this boundedness is to write the closed-loop error dynamics as

$$\dot{z} = f_z(t, z, e_\theta), \qquad (5.36)$$

where f_z denotes the closed-loop function obtained from the open-loop model f combined with the closed-loop feedback. The dynamics (5.36) can be seen as a new state-space model, with states z and input e_θ. Next, an ISS argument can be used for these dynamics, to ensure the boundedness of the states z for bounded input e_θ. After that, we will be able to deduce an IOS property between the extended input $(y_{\mathrm{ref}}^T, e_\theta^T)^T$ and the output z, which eventually, considering the properties assumed on h, will lead to boundedness of the state vector x.

First, to prove ISS between z and input e_θ in Eq. (5.36), we substitute Eq. (5.35) into Eq. (5.31) to obtain

$$\delta = -\tilde{B}^T P z \| L(\xi, t) \|^2 - \hat{E}(t) L(\xi, t) + \Delta b(t, \xi(t)),$$

$$= -\tilde{B}^T P z \| L(\xi, t) \|^2 - \hat{E}(t) L(\xi, t) + E L(\xi, t).$$

After, we consider the same Lyapunov function in Eq. (5.33) as a Lyapunov function for the dynamics (5.36). Then, from Eq. (5.33), we obtain

$$\dot{V} \leq -\|z\|^2 + 2 z^T P \tilde{B} E \, L(\xi, t) - 2 z^T P \tilde{B} \hat{E}(t) L(\xi, t)$$

$$- 2 \| z^T P \tilde{B} \|^2 \| L(\xi, t) \|^2,$$

which leads to

$$\dot{V} \le -\|z\|^2 + 2z^T P\tilde{B}e_E L(\xi, t) - 2\|z^T P\tilde{B}\|^2 \|L(\xi, t)\|^2.$$

Because

$$z^T P\tilde{B}e_E L(\xi) \le \|z^T P\tilde{B}e_E L(\xi)\| \le \|z^T P\tilde{B}\| \|e_E\|_F \|L(\xi)\|$$
$$= \|z^T P\tilde{B}\| \|e_\theta\| \|L(\xi)\|,$$

we can write

$$\dot{V} \le -\|z\|^2 + 2\|z^T P\tilde{B}\| \|e_\theta\| \|L(\xi, t)\| - 2\|z^T P\tilde{B}\|^2 \|L(\xi, t)\|^2,$$

$$\le -\|z\|^2 - 2\left(\|z^T P\tilde{B}\| \|L(\xi, t)\| - \frac{1}{2}\|e_\theta\|\right)^2 + \frac{1}{2}\|e_\theta\|^2,$$

$$\le -\|z\|^2 + \frac{1}{2}\|e_\theta\|^2.$$

Thus we have the following relation:

$$\dot{V} \le -\frac{1}{2}\|z\|^2, \quad \forall \|z\| \ge \|e_\theta\| > 0.$$

Then from Eq. (5.2), we obtain that the system (5.36) is ISS from the input e_θ to the state z. This implies, by Definition 5.1, the following bound on z:

$$\|z(t)\| \le \beta(\|z(t_0)\|, t - t_0) + \gamma\left(\sup_{t_0 \le \tau \le t} \|e_\theta(\tau)\|\right),$$

for any initial state $z(t_0)$, where β is class $\mathcal{KL}$ function and γ is class $\mathcal{K}$. Next, due to the boundedness of the reference trajectory y_{ref}, it is straightforward to extend the previous bound to

$$\|z(t)\| \le \beta(\|z(t_0)\|, t - t_0) + \gamma\left(\sup_{t_0 \le \tau \le t} \|(y_{\text{ref}}^T, e_\theta(\tau)^T)^T\|\right),$$

which by Definition 5.2 implies IOS of the closed-loop system between the input $(y_{\text{ref}}^T, e_\theta(\tau)^T)^T$ and the output z, that is, satisfaction of Assumption 5.7. Finally, boundedness of the state vector x follows by the $\mathcal{K}_\infty$ assumption of h. Now that we found the proper feedback controller that ensures boundedness of the closed-loop system's signals, we can proceed with the parameter identification using similar tools as in the stable open-loop case. Let us summarize the procedure in the following lemma.

Lemma 5.3. *Consider the system (5.19), (5.20), with the uncertain part (5.34), under Assumptions 5.8 and 5.9. Then, if we apply the closed-loop controller (5.23) with Eqs. (5.28), (5.35), we ensure that the norms $\|z\|$, $\|x\|$ are bounded* *for any bounded norm $\|(y_{ref}^T, e_\theta^T)^T\|$, and that the unknown parameters' vector θ, that is, the elements of the matrix E, can be identified using the cost function*

$$Q(\hat{\theta}) = F(e_y(\hat{\theta})), \quad F(0) = 0, \quad F(e_y) > 0, \quad \text{for } e_y \neq 0$$
$$e_y(t) = y(t) - y_{ref}(t),$$
(5.37)

and the estimation algorithm (5.12), (5.13), which under the same assumptions as in Theorem 5.3 leads to the estimation bound (5.14).

Remark 5.10. *Similar to Lemma 5.2, in the practical case of reference trajectories with finite time support $[t_0, t_f]$, here again we can discretize the identification algorithm, that is, piecewise constant evolution of the estimated parameters, and write the corresponding estimation bounds.*

In Case 1, we considered that the uncertain part of the model is a function exactly equal to the unknown parameters' matrix to be identified, multiplied by a known function of time and state variables. Next, we relax this assumption by considering the case where the uncertain term is not exactly known, however, a known upper bound of the uncertainty can be obtained. This can be useful in some mechatronics applications, where the model obtained from the physics often contains trigonometric terms which can be upper bounded by simpler functions.

5.4.2.2 Case 2

We assume here that the uncertain term structure is not exactly known, but is upper bounded by a known function multiplied by an unknown parameters' matrix to be identified, that is,

$$\|\Delta b(t, \xi)\| \leq \|E\|_F \|L(\xi, t)\|,$$
(5.38)

where $L(\xi, t): \mathbb{R}^n \times \mathbb{R} \to \mathbb{R}^m$ is a known bounded function of states and time variables, and $E \in \mathbb{R}^{m \times m}$ is a constant unknown matrix of parameters to be identified. Similar to the previous case, we define the estimation error vector by $e_E = E - \hat{E}$, where $\hat{E}(t)$ denotes the estimate of E. We define the unknown parameter vector as $\theta = [E(1,1), \ldots, E(m,m)]^T \in \mathbb{R}^{m^2}$, and its estimate vector is denoted by $\hat{\theta}(t) = [\hat{E}(1,1), \ldots, \hat{E}(m,m)]^T$, which leads to the estimation error vector $e_\theta(t) = \theta - \hat{\theta}(t)$. Next, we want to write a feedback controller which ensures boundedness of the closed-loop states for bounded identification error and bounded reference trajectory. Similar to the previous case, to ensure that Assumption 5.7 holds, we propose to robustify the nominal controller part (5.23) with the following control term:

$$u_r = -A^{-1}(\xi)\tilde{B}^T Pz\|L(\xi)\|^2 - A^{-1}(\xi)\|\hat{\theta}(t)\|\|L(\xi)\|\text{sign}(\tilde{B}^T Pz).$$
(5.39)

Here again we will use an ISS argument to conclude about the boundedness of the closed–loop system states. To do so, we first write the closed-loop error dynamics as in Eq. (5.36), and prove ISS between e_θ and z as follows: by combining Eqs. (5.31), (5.39) we obtain

$$\delta = -\tilde{B}^T Pz\|L(\xi)\|^2 - \|\hat{\theta}(t)\|\|L(\xi)\|\mathrm{sign}(\tilde{B}^T Pz) + \Delta b(\xi(t)). \quad (5.40)$$

We consider $V(z) = z^T Pz$ as a Lyapunov function candidate for the error dynamics (5.36), where $P > 0$ is a solution of Eq. (5.27).

We can derive that

$$\lambda_{\min}(P)\|z\|^2 \leq V(z) \leq \lambda_{\max}(P)\|z\|^2, \quad (5.41)$$

where $\lambda_{\min}(P) > 0$ and $\lambda_{\max}(P) > 0$ denote the minimum and the maximum eigenvalues of the matrix P, respectively. Then, from Eq. (5.33), we obtain

$$\dot{V} = -\|z\|^2 + 2z^T P\tilde{B}\Delta b(\xi(t)) - 2\|z^T P\tilde{B}\|^2\|L(\xi)\|^2 \\ - 2\|z^T P\tilde{B}\|_1\|\hat{\theta}(t)\|\|L(\xi)\|.$$

Because $\|z^T P\tilde{B}\| \leq \|z^T P\tilde{B}\|_1$, we have

$$\dot{V} \leq -\|z\|^2 + 2z^T P\tilde{B}\Delta b(\xi(t)) - 2\|z^T P\tilde{B}\|^2\|L(\xi)\|^2 \\ - 2\|z^T P\tilde{B}\|\|\hat{\theta}(t)\|\|L(\xi)\|.$$

Then based on the fact that $z^T P\tilde{B}\Delta b(\xi(t)) \leq \|z^T P\tilde{B}\|\|\Delta b(\xi(t))\|$, we obtain

$$\dot{V} \leq -\|z\|^2 + 2\|z^T P\tilde{B}\|\|E\|_F\|L(\xi)\| \\ - 2\|z^T P\tilde{B}\|^2\|L(\xi)\|^2 - 2\|z^T P\tilde{B}\|\|\hat{\theta}(t)\|\|L(\xi)\|, \\ = -\|z\|^2 - 2\|z^T P\tilde{B}\|^2\|L(\xi)\|^2 \\ + 2\|z^T P\tilde{B}\|\|L(\xi)\|(\|\theta\| - \|\hat{\theta}(t)\|).$$

Because $\|\theta\| - \|\hat{\theta}(t)\| \leq \|e_\theta\|$, we can write

$$\dot{V} \leq -\|z\|^2 - 2\|z^T P\tilde{B}\|^2\|L(\xi)\|^2 + 2\|z^T P\tilde{B}\|\|L(\xi)\|\|e_\theta\|.$$

Further, we can obtain

$$\dot{V} \leq -\|z\|^2 - 2\left(\|z^T P\tilde{B}\|\|L(\xi)\| - \frac{1}{2}\|e_\theta\|\right)^2 + \frac{1}{2}\|e_\theta\|^2, \\ \leq -\|z\|^2 + \frac{1}{2}\|e_\theta\|^2.$$

Thus we have the following relation:

$$\dot{V} \leq -\frac{1}{2}\|z\|^2, \quad \forall \|z\| \geq \|e_\theta\| > 0.$$

Then from Eq. (5.2), we obtain that the closed-loop error dynamics (5.36) is ISS from input e_θ to state z. Finally, following similar reasoning as in Lemma 5.3, we can conclude about the boundedness of the closed-loop system states and the convergence of the parameters' estimates.

5.5 IDENTIFICATION AND STABLE PDEs' MODEL REDUCTION BY ES

In this section we will show how to apply the MES-based identification approach to the challenging problem of PDEs' identification and stable model reduction. Indeed, PDEs are valuable mathematical models which are used to describe a large class of systems. For instance, they are used to model fluid dynamics (Rowley, 2005; Li et al., 2013; MacKunis et al., 2011; Cordier et al., 2013; Balajewicz et al., 2013); flexible beams and ropes (Montseny et al., 1997; Barkana, 2014); crowd dynamics (Huges, 2003; Colombo and Rosini, 2005), and so on. However, Being infinite dimension systems, PDEs are almost impossible to solve in closed form (except for some exceptions) and are hard to solve numerically; that is, they require a large computation time. Due to this complexity, it is often hard to use PDEs directly to analyze, predict, or control systems in real time. Instead, one viable approach often used in real applications is to first reduce the PDE model to a "simpler" ODE model, which has a finite dimension, and then use this ODE to analyze, predict, or control the system. The step of obtaining an ODE which represents the original PDE as close as possible is known as model reduction. The obtained ODE is known as the ROM. One of the main problems in model reduction is the identification of some unknown parameters of the ROM or the ODE which also appear in the original PDE, that is, physical parameters of the system. We will call this the ROM identification problem. Another very important problem in PDEs' model reduction is the so-called stable model reduction. Indeed, PDEs are usually reduced by some type of projections (we will explain this step in more detail later in this chapter) into a smaller dimension space. However, the projection step, that is, the dimension reduction step, implies some sort of truncation of the original PDE model, which can lead to instability (in Lagrange stability sense) of the dynamics of the ROM. In other words, the ROM can have radially unbounded solutions, whereas the "true" original

PDE model has bounded solutions. This discrepancy between the ODE ROM and the PDE has to be kept under control or corrected, and this step is usually referred to as ROM stabilization. Many results have been proposed for stable model reduction and ROM parameters' identification problems. In general, these results are presented for some specific PDE, for example, the Navier-Stokes equation, the Boussinesq equation, the Burgers' equation, and so on, representing a particular class of systems, for example, airflow dynamics in a room or fluid motion in a container. For instance, in MacKunis et al. (2011), the authors used a specific model reduction method referred to as proper orthogonal decomposition (POD) to reduce the general nonlinear Navier-Stokes PDE to a system of bilinear ODEs. The authors then used the simplified bilinear ODE model to design a sliding-mode nonlinear estimator to estimate the fluid flow velocity based on some finite velocity field measurements. The convergence of the estimator is shown to be reached in finite time, which is a well-known characteristic of sliding-mode estimators.

In Guay and Hariharan (2008), airflow estimation in buildings has been investigated. Starting from the usual Navier-Stokes PDEs in 2D and in 3D, the authors performed a model reduction using classical POD technique. Then, assuming that the obtained bilinear ODE models are "good enough," that is, stable and with solutions that are close enough to the solutions of the original PDEs' model, the authors used them to develop two types of estimators. First, the authors presented a numerical estimator, based on a simple nonlinear optimization formulation. In this formulation, a nonlinear cost function is defined as a measure of the distance between the measured flow velocities and the estimated flow velocities over a finite time support. Then, this cost function is minimized with respect to the estimated velocities' projection-base coefficients, under dynamical constraints, representing the ODE model as well as algebraic constraints representing the ODE model's output, that is, the estimated flow velocity. This optimization problem was solved numerically to obtain the best estimated velocities' projection-base coefficients, which led to the best flow velocity estimation at any given time within the optimization time support. The authors also considered another estimation technique, namely, the extended Kalman filter, which is a well-known estimation technique for nonlinear ODEs. The authors showed numerically the performance of their estimation on a 2D case and a 3D case.

In Li et al. (2013), the problem of indoor thermal environment modeling and control was studied. Similar to the work mentioned previously, the proposed scheme in this paper uses two stages: an offline stage and an online stage. In the offline stage, the PDE model of the indoor thermal environment coupling temperature distribution with the airflow is reduced to an ODE model using finite volume method (FVM) coupled with classical snapshot-based POD method. One noticeable difference with respect to the previous work is that the obtained ODE model is linear rather than bilinear; this is due to the use of an FVM first before applying a POD reduction. After the obtention of a reduced order ODE model, a Kalman filter was used to estimate the airflow velocity starting from unknown initial conditions and assuming some measurement noises. Eventually, the developed model and estimator were used to develop an MPC-based controller to track a desired reference trajectory. The authors showed the numerical performance of the proposed technique.

In Rowley (2005), the author studied the problem of fluid's PDE reduction to finite dimension ODEs. The author compared three different model reduction methods; POD, balanced truncation, and the balanced POD method. POD, a Galerkin-type projection method, is a well-known model reduction approach that has been extensively used in fluid dynamics model reduction. However, despite it straightforward nature, this method can be sensitive to the choice of the empirical data and the choice of base functions used for the projection. In fact, this method, if not tuned properly, can lead to unstable models even near stable equilibrium points.[1] On the other hand, balanced truncation methods developed mainly in the control theory community for nonlinear systems do not suffer from the sensitivity drawbacks of POD methods. Unfortunately, balanced truncation methods suffer from the so-called curse of dimensionality because they are computationally demanding and do not scale well with the dimension of the dynamics to be reduced, which makes them less practical for fluid systems applications. The third method, balanced POD, combines ideas from POD and balanced truncation. This method uses empirical Gramians to compute balanced truncations for nonlinear systems for systems with very large dimensions. The three methods were then compared on a linearized flow in a plane channel.

In Balajewicz et al. (2013), a new model reduction technique for Navier-Stokes equations was proposed. The new method is based on the constatation that the classical POD approach uses basis functions that model only large energy flow dynamics, However, for the Navier-Stokes equations

---

[1] Hereafter, by stability we mean the stability of flows' solutions in the sense of Lagrange.

with high Reynolds number the flow dynamics are characterized with both large and low energy scales. Because POD fails to capture low energy scales of the solutions, it does not reproduce accurately the solutions of the original dynamics. The authors in Balajewicz et al. (2013), introduced a new idea in which they use both high and low energy scale basis functions. To do so they formulated the basis function selection problem as a problem of tuning the kinetic energy of a spectrally discretized fluid flow. Indeed, knowing that a positive rate of kinetic energy production is associated with basis functions that reproduce a greater proportion of large energy scales, and that a negative rate of kinetic energy production is associated with basis functions that resolve small energy scales, one could find "optimal" basis functions that capture small and large energy scales by forcing the rate of kinetic energy to be positive or negative. The authors formulated this "tuning" problem as a finite dimensional optimization problem, which led to optimal basis functions.

In the following, we propose to treat the problem of ROM physical parameters' identification and the problem of ROM stabilization in two separate sections.

5.5.1 ES-Based ROM Parameters' Identification

Consider a stable dynamical system modeled by a nonlinear PDE of the form

$$\dot{z} = \mathcal{F}(z, p) \in \mathcal{Z}, \tag{5.42}$$

where $\mathcal{Z}$ is an infinite-dimension Hilbert space, and $p \in \mathbb{R}^m$ represents the vector of physical parameters to be identified. While solutions to this PDE can be obtained through numerical discretization, for example, finite elements, finite volumes, finite differences, and so on; these computations are often very expensive and not suitable for online applications, for example, airflow analysis, prediction, and control. However, solutions of the original PDE often exhibit low rank representations in an "optimal" basis, which is exploited to reduce the PDE to a finite dimension ODE.

The general idea is as follows: One first finds a set of "optimal" (spatial) basis vectors $\phi_i \in \mathbb{R}^n$ (the dimension n is generally very large and comes from a "brute-force" discretization of the PDE, e.g., finite element discretization), and then approximates the PDE solution as

$$z(t) \approx \Phi z_r(t) = \sum_{i=1}^{r} z_{ri}(t) \phi_i, \tag{5.43}$$

where Φ is an $n \times r$ matrix containing the basis vectors ϕ_i as column vectors. Next, the PDE is projected into the finite r-dimensional space via classical nonlinear model reduction techniques, for example, Galerkin projection, to obtain an ROM of the form

$$\dot{z}_r(t) = F(z_r(t), p) \in \mathbb{R}^r, \tag{5.44}$$

where $F: \mathbb{R}^r \rightarrow \mathbb{R}^r$ is obtained from the original PDE structure through the model reduction technique, for example, the Galerkin projection. Clearly, the problem lies in the selection of this "optimal" basis matrix Φ. There are many model reduction methods to find the projection basis functions for nonlinear systems. For example, POD, dynamic mode decomposition (DMD), and reduced basis are some of the most used methods. We will recall hereafter the POD method; however, the ES-based identification results are independent of the type of model reduction approach, and the results of this chapter remain valid regardless of the selected model reduction method.

5.5.1.1 POD Basis Functions

We give here a brief recall of POD basis function computation; the interested reader can refer to Kunisch and Volkwein (2007) and Gunzburger et al. (2007) for a more complete exposition about POD theory.

The general idea behind POD is to select a set of basis functions that capture an optimal amount of energy of the original PDE. The POD basis functions are obtained from a collection of snapshots over a finite time support of the PDE solutions. These snapshots are usually obtained by solving an approximation (discretization) of the PDE, for example, using finite element method (FEM), or by direct measurements from the system which is modeled by the PDE, if real measurements are available. The POD basis function computation steps are presented next in more detail.

We consider here the case where the POD basis functions are computed mathematically from approximation solution snapshots of the PDE (no real measurements are used). First, the original PDE is discretized using any finite element basis functions, for example, piecewise linear functions or spline functions, and so on (we are not presenting here any FEM; instead, we refer the reader to the numerous manuscripts in the field of FEM, for example, Sordalen, 1997; Fletcher, 1983). Let us denote the associated PDE solutions' approximation by $z_{\text{fem}}(t, x)$, where t stands for the scalar time variable, and x stands for the space variable, often referred to as the dimension of the PDE; that is, x is scalar for one dimension and a vector of two elements in a two-dimensional space, and so on. We consider here (for simplicity of the notations) the case of one dimension where x is a scalar in a finite interval, which we consider without loss of generality to be $[0, 1]$. Next, we compute a set of s snapshots of approximation solutions as

$$S_z = \{z_{\text{fem}}(t_1, \cdot), \ldots, z_{\text{fem}}(t_s, \cdot)\} \subset \mathbb{R}^N, \tag{5.45}$$

where N is the selected number of FEM basis functions.

We define the so-called correlation matrix K^z elements as

$$K^z{}_{ij} = \frac{1}{s}\langle z_{\text{fem}}(t_i, \cdot), z_{\text{fem}}(t_j, \cdot)\rangle, \quad i, j = 1, \ldots, s. \tag{5.46}$$

We then compute the normalized eigenvalues and eigenvectors of K^z, denoted as λ^z, and v^z (note that λ^z are also referred to as the POD eigenvalues). Eventually, the ith POD basis function is given by

$$\phi_i^{\text{pod}}(x) = \frac{1}{\sqrt{s}\sqrt{\lambda_i^z}} \sum_{j=1}^{j=s} v_i^z(j) z^{\text{fem}}(t_j, x), \quad i = 1, \ldots, N_{\text{pod}} \tag{5.47}$$

where $N_{\text{pod}} \leq s$ is the number of retained POD basis functions, which depends on the application.

One of the main properties of the POD basis functions is orthonormality, which means that the basis functions satisfy the following equalities:

$$\langle \phi_i^{\text{pod}}, \phi_j^{\text{pod}} \rangle = \int_0^1 \phi_i^{\text{pod}}(x)\phi_j^{\text{pod}}(x)\,dx = \delta_{ij}, \tag{5.48}$$

where δ_{ij} denotes the Kronecker delta function. The solution of the PDE (5.42) can then be approximated as

$$z^{\text{pod}}(t, x) = \sum_{i=1}^{i=N_{\text{pod}}} \phi_i^{\text{pod}}(x) q_i^{\text{pod}}(t), \tag{5.49}$$

where $q_i^{\text{pod}}, i = 1, \ldots, N_{\text{pod}}$ are the POD projection coefficients (which play the role of the z's in the ROM (5.44)). Finally, the PDE (5.42) is projected on the reduced dimension POD space using a Galerkin projection; that is, both sides of Eq. (5.42) are multiplied by the POD basis functions, where z is substituted by z^{pod}, and then both sides are integrated over the space interval $[0, 1]$. Finally, using the orthonormality constraints (5.48) and the boundary constraints of the original PDE, one obtains an ODE of the form

$$\dot{q}^{\text{pod}}(t) = F(q^{\text{pod}}(t), p) \in \mathbb{R}^{N_{\text{pod}}}, \tag{5.50}$$

where the structure (in terms of nonlinearities) of the vector field F is related to the structure of the original PDE, and where $p \in \mathbb{R}^m$ represents the vector of parametric uncertainties to be identified.

We can now proceed with the MES-based estimation of the parametric uncertainties.

5.5.1.2 MES-Based Open-Loop Parameters' Estimation for PDEs

Similar to Section 5.3.2, we will use MES to estimate the PDEs' parametric uncertainties using its ROM, that is, the POD ROM. First, we need to introduce some basic stability assumptions.

Assumption 5.12. *The solutions of the original PDE model (5.42) are assumed to be in $L^2([0, \infty); \mathcal{Z})$, and the associated POD ROM (5.49), (5.50) is Lagrange stable.*

Remark 5.11. *Assumption 5.12 is needed to be able to perform open-loop identification of the system. The case of unstable ROMs will be studied in the next section.*

To be able to use the MES framework to identify the parameters' vector p, we define an identification cost function as

$$Q(\hat{p}) = H(e_z(\hat{p})), \tag{5.51}$$

where $\hat{p}$ denotes the estimate of p, H is a positive definite function of e_z, and e_z represents the error between the ROM (5.49), (5.50) and the system's measurements z_m, defined as

$$e_z(t) = z^{\text{pod}}(t, x_m) - z_m(t, x_m), \tag{5.52}$$

x_m being the points in space where the measurements are obtained.

To formulate an upper bound on the estimation error norm, we add the following assumptions on the cost function Q.

Assumption 5.13. *The cost function Q has a local minimum at $\hat{p}^* = p$.*

Assumption 5.14. *The original parameters' estimates vector $\hat{p}$; that is, the vector of nominal parameters is close enough to the actual parameters' vector p.*

Assumption 5.15. *The cost function is analytic and its variation with respect to the uncertain variables is bounded in the neighborhood of p^*, that is, $\|\frac{\partial Q}{\partial p}(\tilde{p})\| \leq \xi_2, \xi_2 > 0, \tilde{p} \in \mathcal{V}(p^*)$, where $\mathcal{V}(p^*)$ denotes a compact neighborhood of p^*.*

Based on the previous assumptions we can summarize the ODE open-loop identification result in the following lemma.

Lemma 5.4. *Consider the system (5.42), under Assumptions 5.12–5.15. The vector of uncertain parameters p can be estimated online using the algorithm*

$$\hat{p}(t) = p_{nom} + \Delta p(t), \tag{5.53}$$

where p_{nom} is the nominal value of p, and $\Delta p = [\delta p_1, \ldots, \delta p_m]^T$ is computed using the MES algorithm

$$
\dot{z}_i = a_i \sin\left(\omega_i t + \frac{\pi}{2}\right) Q(\hat{p}),
$$
$$
\delta p_i = z_i + a_i \sin\left(\omega_i t - \frac{\pi}{2}\right), \quad i \in \{1, \ldots, m\}
\tag{5.54}
$$

with $\omega_i \neq \omega_j, \omega_i + \omega_j \neq \omega_k, i, j, k \in \{1, \ldots, m\}$, and $\omega_i > \omega^*, \forall i \in \{1, \ldots, m\}$, with ω^* large enough, and Q given by Eqs. (5.51), (5.52), with the estimate upper bound

$$
\|e_p(t)\| = \|\hat{p} - p\| \leq \frac{\xi_1}{\omega_0} + \sqrt{\sum_{i=1}^{i=m} a_i^2}, \quad t \to \infty
\tag{5.55}
$$

where $\xi_1 > 0$ and $\omega_0 = \max_{i \in \{1, \ldots, m\}} \omega_i$.

The proof follows the same steps as in the proof of Theorem 5.2.

Similar to the results in Section 5.3, we should consider the "more practical" case where the estimation is done over a finite time support. This case is summarized in the following lemma.

Lemma 5.5. *Consider the system (5.42), over a finite time support, that is, $t \in [t_0, t_f], 0 \leq t_0 < t_f$. Then, the parameter vector p can be iteratively estimated starting from its nominal value p_{nom}, such that*

$$
\hat{p}(t) = p_{nom} + \Delta p(t),
$$
$$
\Delta p(t) = \hat{\Delta} p((I - 1)t_f), \quad (I - 1)t_f \leq t < It_f, \quad I = 1, 2, \ldots
\tag{5.56}
$$

where I is the iteration number and $\hat{\Delta} p = [\hat{\delta} p_1, \ldots, \hat{\delta} p_m]^T$ is computed using the iterative MES algorithm

$$
\dot{z}_i = a_i \sin\left(\omega_i t + \frac{\pi}{2}\right) Q(\hat{p}),
$$
$$
\hat{\delta} p_i = z_i + a_i \sin\left(\omega_i t - \frac{\pi}{2}\right), \quad i \in \{1, \ldots, m\}
\tag{5.57}
$$

with Q given by Eqs. (5.51), (5.52) $\omega_i \neq \omega_j, \omega_i + \omega_j \neq \omega_k, i, j, k \in \{1, \ldots, m\}$, and $\omega_i > \omega^*, \forall i \in \{1, \ldots, m\}$, with ω^* large enough. Then under Assumptions 5.12–5.15, the estimate upper bound is given by

$$
\|\hat{p}(It_f) - p\| \leq \frac{\xi_1}{\omega_0} + \sqrt{\sum_{i=1}^{i=m} a_i^2}, \quad I \to \infty
$$

where $\xi_1 > 0$ and $\omega_0 = \max_{i \in \{1, \ldots, m\}} \omega_i$.

So far we have assumed that the PDE solutions as well as the ROM solutions satisfy Assumption 5.12. However, in some cases, even if the PDE admits Lagrange stable solutions, the associated ROM can exhibit radially unbounded solutions over time (e.g., refer to Cordier et al. (2013) for the Navier-Stokes POD ROM instability case). This phenomenon is well known in the fluid mechanics community, where some ROMs are stable for some boundary conditions of the PDE and then become unstable when the boundary conditions change. For example, it is well known in the fluid dynamics community, see, Noack et al. (2011), Ahuja and Rowley (2010), and Barbagallo et al. (2009), that for laminar fluid flows the Galerkin-based POD ROM (POD-ROM-G) leads to stable solutions. However, for turbulent flows, see, Aubry et al. (1988) and Wang et al. (2012), the POD-ROM-G might lead to unstable solutions. One explanation for this phenomenon is that some of the high index POD modes which are truncated in the Galerkin projection step can have an important role in reproducing the stable behavior of the actual flow.

To remedy this instability problem of ROMs, the fluid dynamics community proposed a large repertoire of the so-called "closure models." The term closure models refers to an additional linear or nonlinear term added to the ROM to recover the "physical" stabilizing effect of the truncated modes. For example, in Aubry et al. (1988), a POD ROM has been stabilized using an eddy-viscosity term to model the effect of truncated POD modes for the problem of turbulent boundary layers. In the paper, San and Borggaard (2014), the authors study the challenging case of the unsteady Marsigli flow problem with Kelvin-Helmholtz instability (induced by temperature jumps), using POD ROM stabilization of the Boussinesq equations. Two eddy viscosity closure models were used to recover the dissipating (stabilizing) effect of the truncated POD modes. The first closure model is based on the addition of a constant eddy viscosity coefficient to the POD ROM; this method is also known as the Heisenberg stabilization, see for example, Bergmann et al. (2009). The second closure model (introduced by Rempfer, 1991) is based on a different constant eddy viscosity coefficient for each mode. Other related closure model extensions were proposed and tested on the Burgers' equation in San and Iliescu (2013). Many other closure model studies have been published, see, Wang (2012), Noack et al. (2003, 2005, 2008), Ma and Karniadakis (2002), and Sirisup and Karniadakis (2004). We do not pretend here to present all these results in detail; instead, we will focus in the next section on some of the models proposed in San and Iliescu (2013) and Cordier et al. (2013), and use them to demonstrate the utility of MES algorithms in tuning closure models for PDEs' model reduction and stabilization.

5.5.2 MES-Based PDEs' Stable Model Reduction

Let us first present the problem of stable model reduction in its general form, that is, without specifying a particular type of PDE. If we consider the general PDE given by Eq. (5.42), where the parameter p is substituted by the physical parameter representing the viscosity denoted by μ, we can rewrite the PDE model as

$$\dot{z} = \mathcal{F}(z, \mu) \in \mathcal{Z}, \quad \mu \in \mathbb{R}. \tag{5.58}$$

Next, we apply the POD model reduction technique to the model (5.58), leading to an ODE model of the form (5.49), (5.50), which writes as

$$\begin{cases} \dot{q}^{\text{pod}}(t) = F(q^{\text{pod}}(t), \mu), \\ z^{\text{pod}}(t, x) = \sum_{i=1}^{i=N_{\text{pod}}} \phi_i^{\text{pod}}(x) q_i^{\text{pod}}(t). \end{cases} \tag{5.59}$$

As we explained earlier, the problem with this "simple" Galerkin POD ROM (denoted POD ROM-G) is that the norm of z^{pod} might become unbounded over a given time support, whereas z, the solution of Eq. (5.58), is actually bounded. One of the main ideas behind the closure model approach is that the viscosity coefficient μ in Eq. (5.59) is substituted by a virtual viscosity coefficient μ_{cl}, the form of which is chosen in such a way to stabilize the solutions of the POD ROM (5.59). Another idea is to add a penalty term H to the original (POD) ROM-G, as follows:

$$\dot{q}^{\text{pod}}(t) = F(q^{\text{pod}}(t), \mu) + H(t, q^{\text{pod}}). \tag{5.60}$$

The structure of the term H is chosen depending on the structure of F, in such a way to stabilize the solutions of Eq. (5.59), for example, the Cazemier penalty model reported in San and Iliescu (2013). We will recall different closure models presented in San and Iliescu (2013) and Cordier et al. (2013).

   Remark 5.12. *We want to underline here that although we are presenting the idea of closure models in the framework of POD ROM, it is not limited to ROMs obtained by POD techniques. Indeed, closure model ideas, at least in a mathematical sense, can be applied to ROMs obtained by other model-reduction techniques, for example, DMD.*

5.5.2.1 Different Closure Models for ROM Stabilization

We first recall here several closure models introduced in San and Iliescu (2013) for the case of Burgers' equations. We will still maintain a general presentation here, even though the models in San and Iliescu (2013) have been tested on the Burgers' equations, because similar closure models could be used on other PDEs.

(A) Closure models with constant eddy viscosity coefficients
-   An eddy viscosity model known as the Heisenberg ROM (ROM-H) is simply given by the constant viscosity coefficient

$$\mu_{cl} = \mu + \mu_e, \tag{5.61}$$

where μ is the nominal value of the viscosity coefficient in Eq. (5.58), and μ_e is the additional constant term added to compensate for the damping effect of the truncated modes.
-   A variation of the ROM-H, which we will denote as ROM-R (because it was proposed in Rempfer, 1991), is where μ_e is dependent on the mode index, and where the viscosity coefficients (for each mode) are given by

$$\mu_{cl} = \mu + \mu_e \frac{i}{R}, \tag{5.62}$$

with μ_e being the viscosity amplitude, i the mode index, and R the number of retained modes in the ROM computation.
-   Another model proposed in San and Iliescu (2013) is somehow a quadratic version of the ROM-R, and so is denoted as ROM-RQ. It is given by the coefficients

$$\mu_{cl} = \mu + \mu_e \left(\frac{i}{R}\right)^2, \tag{5.63}$$

where the variables are defined similarly as in Eq. (5.62).
-   A model proposed in San and Iliescu (2013) is a root-square version of the ROM-R, and is denoted as ROM-RS. It is given by

$$\mu_{cl} = \mu + \mu_e \sqrt{\frac{i}{R}}, \tag{5.64}$$

where the coefficients are defined as in Eq. (5.62).
-   Spectral vanishing viscosity models are similar to the ROM-R in the sense that the amount of induced damping changes as a function of the mode index. This concept has been introduced in Tadmor (1989), and so these closure models are denoted as ROM-T, and are given by

$$\begin{cases} \mu_{cl} = \mu, & \text{for } i \leq M, \\ \mu_{cl} = \mu + \mu_e, & \text{for } i > M, \end{cases} \tag{5.65}$$

where i denotes the mode index and $M \leq R$ is the index of modes above which a nonzero damping is introduced, R being the total number of ROM modes.

- A model introduced in Sirisup and Karniadakis (2004) falls into the class of vanishing viscosity models; we denote it as ROM-SK, given by

$$
\begin{cases}
\mu_{\mathrm{cl}} = \mu + \mu_e e^{\frac{-(i-R)^2}{(i-M)^2}} & \text{for } i \leq M, \\
\mu_{\mathrm{cl}} = \mu & \text{for } i > M, M \leq R.
\end{cases}
\tag{5.66}
$$

- The last model has been introduced in Chollet (1984) and Lesieur and Metais (1996), is denoted as ROM-CLM, and is given by

$$
\mu_{\mathrm{cl}} = \mu + \mu_e \alpha_0^{-1.5} \left(\alpha_1 + \alpha_2 e^{-\frac{\alpha_3 R}{i}} \right),
\tag{5.67}
$$

where i is the mode index, $\alpha_0, \alpha_1, \alpha_2, \alpha_3$ are positive gains (see Karamanos and Karniadakis, 2000; Chollet, 1984 for some insight about their tuning), and R is the total number of the ROM modes.

(B) Closure models with time- and space-varying eddy viscosity coefficients

Several varying (in time and/or space) viscosity terms have been proposed in the literature. For instance, in San and Iliescu (2013), the Smagorinsky nonlinear viscosity model has been reported. However, the Smagorinsky model is based on the online computation of some nonlinear terms at each time step, which in general makes them computationally consuming. We report here the nonlinear viscosity model presented in Cordier et al. (2013), which is nonlinear and a function of the ROM state variables. To be able to report this model, we first need to rewrite the ROM (5.59) in such a way to display explicitly the linear viscous term, as follows:

$$
\begin{cases}
\dot{q}^{\mathrm{pod}}(t) = F(q^{\mathrm{pod}}(t), \mu) = \tilde{F}(q^{\mathrm{pod}}(t)) + \mu \, D q^{\mathrm{pod}}, \\
z^{\mathrm{pod}}(t, x) = \sum_{i=1}^{i=N_{\mathrm{pod}}} \phi_i^{\mathrm{pod}}(x) q_i^{\mathrm{pod}}(t),
\end{cases}
\tag{5.68}
$$

where $D \in \mathbb{R}^{N_{\mathrm{pod}} \times N_{\mathrm{pod}}}$ represents a constant viscosity damping matrix, and the term $\tilde{F}$ represents the rest of the ROM, that is, the part without damping.

Based on Eq. (5.68), we can write the nonlinear eddy viscosity model denoted H_{nev}, as

$$
H_{\mathrm{nev}} = \mu_e \sqrt{\frac{V(q^{\mathrm{pod}})}{V_\infty}} \, \mathrm{diag}(d_{11}, \ldots, d_{N_{\mathrm{pod}} N_{\mathrm{pod}}}) q^{\mathrm{pod}},
\tag{5.69}
$$

where $\mu_e > 0$ is the amplitude of the model, the $d_{ii}, i = 1, \ldots, N_{pod}$ are the diagonal elements of the matrix D, and $V(q^{pod})$, V_∞ are defined as follows:

$$V = \sum_{i=1}^{i=N_{pod}} 0.5 q_i^{pod^2}, \tag{5.70}$$

$$V_\infty = \sum_{i=1}^{i=N_{pod}} 0.5 \lambda_i, \tag{5.71}$$

where the λ_is are the selected POD eigenvalues (as defined in Section 5.5.1.1). We point out here that compared to the previous closure models, the nonlinear closure model H_{nev} is not added to the viscosity term, but rather added directly to the right-hand side of the ROM Eq. (5.68), as an additive stabilizing nonlinear term. The stabilizing effect has been analyzed in Cordier et al. (2013) based on the decrease over time of the energy function $K(t)$ along the trajectories of the ROM solutions, that is, a Lyapunov-type analysis.

We reported in this section several closure models proposed in the community to stabilize ROMs. However, one of the big challenges, see, Cordier et al. (2013) and San and Borggaard (2014), with these closure models is the tuning of their free parameters, for example, the gain μ_e. In the next section we show how MES can be used to auto-tune the closure models' free coefficients to optimize their stabilizing effect.

5.5.2.2 MES-Based Closure Models' Auto-Tuning

As mentioned in San and Borggaard (2014), the tuning of the closure model amplitude is important to achieve an optimal stabilization of the ROM. We will use the model-free MES optimization algorithms to tune the coefficients of the closure models presented in Section 5.5.2.1. The advantage of using MES is the auto-tuning capability that these type of algorithms will allow. Moreover, one important point that we want to highlight here is the fact that the use of MES allows us to constantly tune the closure model, even in an online operation of the system. Indeed, MES can be used offline to tune the closure model, but it can also be connected online to the real system to continuously fine-tune the closure model coefficients, which will make the closure model valid for a longer time interval compared to the classical closure models, which are usually tuned offline over a fixed finite time interval.

Similar to the parameters' identification case, we should first define a suitable learning cost function. The goal of the learning (or tuning) is to enforce Lagrange stability of the ROM (5.59), and to make sure that the solutions of the ROM (5.59) are close to the ones of the original PDE (5.58). The later learning goal is important because the whole process of model reduction aims at obtaining a simplified ODE model which reproduces the solutions of the original PDE (the real system) with much less computation burden.

We define the learning cost as a positive definite function of the norm of the error between the solutions of Eq. (5.58) and the ROM (5.59), as follows:

$$Q(\hat{\mu}) = H(e_z(\hat{\mu})),$$
$$e_z(t) = z^{\text{pod}}(t, x) - z(t, x), \tag{5.72}$$

where $\hat{\mu} \in \mathbb{R}$ denotes the learned parameter μ. Note that the error e_z could be computed in an offline setting where the ROM stabilization problem is solved numerically, based on solutions of the ROM (5.59), and solutions of the PDE (5.58) at a vector of space points x. The error could be also computed online where the z^{pod} is obtained from solving the model (5.59), but the z is obtained from real measurements of the system at selected space points x. Another more practical way of implementing the ES-based tuning of μ is to start with an offline tuning and then use the obtained ROM, that is, the obtained optimal value of μ in an online operation of the system, for example, control and estimation, and then fine-tune the ROM online by continuously learning the best value of μ at any given time during the operation of the system.

To write some formal convergence results, we need some classical assumptions on the solutions of the original PDE, and on the learning cost function.

Assumption 5.16. *The solutions of the original PDE model (5.58) are assumed to be in $L^2([0,\infty); \mathcal{Z})$.*

Assumption 5.17. *The cost function Q in Eq. (5.72) has a local minimum at $\hat{\mu} = \mu^*$.*

Assumption 5.18. *The cost function Q in Eq. (5.72) is analytic and its variation with respect to μ is bounded in the neighborhood of μ^*, that is, $\|\frac{\partial Q}{\partial \mu}(\tilde{\mu})\| \leq \xi_2, \xi_2 > 0, \tilde{\mu} \in \mathcal{V}(\mu^*)$, where $\mathcal{V}(\mu^*)$ denotes a compact neighborhood of μ^*.*

We can now write the following lemma.

Lemma 5.6. *Consider the PDE (5.58), under Assumption 5.16, together with its ROM (5.59), where the viscosity coefficient μ is substituted by μ_{cl}. Then, if μ_{cl} takes the form of any of the closure models in Eqs. (5.61)–(5.67), where the closure model amplitude μ_e is tuned based on the following ES algorithm*

$$
\begin{aligned}
\dot{y} &= a \sin\left(\omega t + \frac{\pi}{2}\right) Q(\hat{\mu}_e), \\
\hat{\mu}_e &= y + a \sin\left(\omega t - \frac{\pi}{2}\right),
\end{aligned}
\tag{5.73}
$$

where $\omega > \omega^*$, ω^* large enough, and Q is given by Eq. (5.72), under Assumptions 5.17 and 5.18, the norm of the distance with respect to the optimal value of μ_e, $e_\mu = \mu^* - \hat{\mu}_e(t)$ admits the following bound:

$$
|e_\mu(t)| \leq \frac{\xi_1}{\omega} + a, \quad t \to \infty,
\tag{5.74}
$$

where $a > 0, \xi_1 > 0$, and the learning cost function approaches its optimal value with the following upper bound:

$$
|Q(\hat{\mu}_e) - Q(\mu^*)| \leq \xi_2 \left(\frac{\xi_1}{\omega} + a\right), \quad t \to \infty,
\tag{5.75}
$$

where $\xi_2 = \max_\mu \in \mathcal{V}(\mu^*) |\frac{\partial Q}{\partial \mu}|$.

Proof. Based on Assumptions 5.17 and 5.18, the ES nonlinear dynamics (5.73) can be approximated by a linear averaged dynamic (using averaging approximation over time, Rotea, 2000a, p. 435, Definition 1). Furthermore, $\exists \; \xi_1, \omega^*$, such that for all $\omega > \omega^*$, the solution of the averaged model $\hat{\mu}_{aver}(t)$ is locally close to the solution of the original ES dynamics, and satisfies (Rotea, 2000a, p. 436)

$$
|\hat{\mu}_e(t) - d(t) - \hat{\mu}_{aver}(t)| \leq \frac{\xi_1}{\omega}, \quad \xi_1 > 0, \qquad \forall t \geq 0
$$

with $d(t) = a \sin(\omega t - \frac{\pi}{2})$. Moreover, because Q is analytic it can be approximated locally in $\mathcal{V}(\mu^*)$ with a quadratic function, for example, Taylor series up to second order, which leads to (Rotea, 2000a, p. 437)

$$
\lim_{t \to \infty} \hat{\mu}_{aver}(t) = \mu^*.
$$

Based on this, we can write

$$
|\hat{\mu}_e(t) - \mu^*| - |d(t)| \leq |\hat{\mu}_e(t) - \mu^* - d(t)| \leq \frac{\xi_1}{\omega}, \quad \xi_1 > 0, \qquad t \to \infty
$$

$$
\Rightarrow |\hat{\mu}_e(t) - \mu^*| \leq \frac{\xi_1}{\omega} + |d(t)|, \quad t \to \infty
$$

which implies

$$|\hat{\mu}_e(t) - \mu^*| \leq \frac{\xi_1}{\omega} + a, \quad \xi_1 > 0, \qquad t \to \infty.$$

Next, the cost function upper bound is easily obtained from the previous bound, using the fact that Q is locally Lipschitz with the Lipschitz constant $\xi_2 = \max_\mu \in \mathcal{V}(\mu^*)|\frac{\partial Q}{\partial \mu}|$.

The closure models based on constant eddy viscosity coefficients can be a good solution to stabilize ROMs and preserve the intrinsic energy properties of the original PDE in some cases where the influence of the linear terms of the PDE are dominant, for example, in short timescales. However, in many cases with nonlinear energy cascade these closure models are unrealistic because linear terms cannot recover the nonlinear energy terms lost during the ROM computation. For this reason many researchers have tried to come up with nonlinear stabilizing terms. The closure model given by Eq. (5.69) is one example of nonlinear closure models which have been proposed in Noack et al. (2008) based on finite-time thermodynamics (FIT) arguments, and in Noack et al. (2011) based on scaling arguments.

Based on this, we propose here to use a combination of both linear and nonlinear closure models. Indeed, we think that the combination of both models can lead to a more efficient closure model which can handle efficiently linear energy terms that might be dominant for small timescales and handle nonlinear energy terms which might be more dominant for large timescales and in some specific PDEs/boundary conditions. Furthermore, we propose to auto-tune these "new" closure models using MES algorithms, which gives an automatic way to select the appropriate term to amplify (even online, if the MES is implemented online as explained earlier in this chapter) either the linear part or the nonlinear part of the closure model, depending on the present behavior of the system, for example, depending on the boundary conditions.

We summarize this result in the following lemma.

Lemma 5.7. *Consider the PDE (5.58), under Assumption 5.16, together with its stabilized ROM*

$$\begin{cases} \dot{q}^{pod}(t) &= F(q^{pod}(t), \mu) = \tilde{F}(q^{pod}(t)) + \mu_{lin} \, Dq^{pod} + H_{nl}(q^{pod}, \mu_{nl}), \\ z^{pod}(t, x) &= \sum_{i=1}^{i=N_{pod}} \phi_i^{pod}(x) q_i^{pod}(t), \\ H_{nl} &= \mu_{nl}\sqrt{\frac{V(q^{pod})}{V_\infty}} \, diag(d_{11}, \ldots, d_{N_{pod}N_{pod}}) q^{pod}, \\ V &= \sum_{i=1}^{i=N_{pod}} 0.5 q_i^{pod2}, \\ V_\infty &= \sum_{i=1}^{i=N_{pod}} 0.5 \lambda_i, \end{cases}$$

$$(5.76)$$

where the linear viscosity coefficient μ_{lin} is substituted by μ_{cl} chosen from any of the constant closure models (5.61), (5.62), (5.63), (5.64), (5.65), (5.66), or (5.67), and where the closure model amplitudes μ_e, μ_{nl} are tuned based on the following MES algorithm:

$$\dot{y}_1 = a_1 \, \sin\left(\omega_1 t + \frac{\pi}{2}\right) Q(\hat{\mu}_e, \hat{\mu}_{nl}),$$

$$\hat{\mu}_e = y_1 + a_1 \, \sin\left(\omega_1 t - \frac{\pi}{2}\right),$$

$$\dot{y}_2 = a_2 \, \sin\left(\omega_2 t + \frac{\pi}{2}\right) Q(\hat{\mu}_e, \hat{\mu}_{nl}), \qquad (5.77)$$

$$\hat{\mu}_{nl} = y_2 + a_2 \, \sin\left(\omega_2 t - \frac{\pi}{2}\right),$$

where $\omega_{\max} = \max(\omega_1, \omega_2) > \omega^*$, ω^* large enough, and Q is given by Eq. (5.72), with $\hat{\mu} = (\hat{\mu}_e, \hat{\mu}_{nl})$, under Assumptions 5.17 and 5.18, the norm of the vector of the distance with respect to the optimal values of μ_e, μ_{nl}; $e_\mu = (\mu_e{}^* - \hat{\mu}_e(t), \mu_{nl}{}^* - \hat{\mu}_{nl}(t))$ admits the following bound:

$$\|e_\mu(t)\| \leq \frac{\xi_1}{\omega_{\max}} + \sqrt{a_1^2 + a_2^2}, \quad t \to \infty, \qquad (5.78)$$

where $a_1, a_2 > 0, \xi_1 > 0$, and the learning cost function approaches its optimal value within the following upper bound:

$$|Q(\hat{\mu}_e, \hat{\mu}_{nl}) - Q(\mu_e{}^*, \mu_{nl}{}^*)| \leq \xi_2 \left(\frac{\xi_1}{\omega} + \sqrt{a_1^2 + a_2^2}\right), \quad t \to \infty \qquad (5.79)$$

where $\xi_2 = \max_{(\mu_1, \mu_2)} \in \mathcal{V}(\mu^*) \|\frac{\partial Q}{\partial \mu}\|$.

Proof. We will skip the proof for this lemma because it follows the same steps as the proof of Lemma 5.6.

Remark 5.13. *What we want to underline here is that having two tuning eddy coefficient amplitudes μ_{lin} and μ_{nl} gives an extra degree of freedom, compared to having only a linear closure term or only a nonlinear closure term. The MES can then choose which term to emphasize, that is, have higher value of its amplitude, depending on the PDE and the boundary conditions under consideration.*

Remark 5.14. *For clarity and generality of the presentation, we kept the PDE and the associated ROM terms very general, that is, the terms $\mathcal{F}$ and $\tilde{F}$. However, we remind the reader at this stage that the nonlinear closure model given by Eq. (5.69) has been proposed in Noack et al. (2008, 2011) for the special case of Navier-Stokes PDE. This means that there is no guarantee that it will work if it is used on a different PDE with different nonlinearities' structure, and we do not mean to state otherwise by Lemma 5.7.*

We want now to consider the general case where $\tilde{F}$ can be an unknown bounded nonlinear function, for example, a function with bounded structured uncertainties. In this case, we will use Lyapunov theory to propose a nonlinear closure model which stabilizes the ROM, and then complement it with an MES learning algorithm to optimize the performance of the ROM in terms of reproducing the original PDE's solutions.

Let us consider again the PDE (5.58) together with its ROM (5.68). We assume that $\tilde{F}$ satisfies the following assumption.

Assumption 5.19. *The norm of the vector field $\tilde{F}$ is bounded by a known function of q^{pod}, that is, $\|\tilde{F}(q^{pod})\| \leq \tilde{f}(q^{pod})$.*

Remark 5.15. *Assumption 5.19 allows us to consider general structures of PDEs and their associated ROMs. Indeed, all we need is that the ROM right-hand side has a structure similar to Eq. (5.68), where an explicit damping linear term is extracted and added to a bounded nonlinear term $\tilde{F}$, which can include any structured uncertainty of the ROM; for example, bounded parametric uncertainties can be formulated in this manner.*

We can now write the following result.

Theorem 5.4. *Consider the PDE (5.58), under Assumption 5.16, together with its stabilized ROM*

$$
\begin{cases}
\dot{q}^{pod}(t) = \tilde{F}(q^{pod}(t)) + \mu\, Dq^{pod} + H_{nl}(q^{pod}), \\
z^{pod}(t, x) = \sum_{i=1}^{i=N_{pod}} \phi_i^{pod}(x) q_i^{pod}(t),
\end{cases}
\tag{5.80}
$$

where $\tilde{F}$ satisfies Assumption 5.19, the diagonal elements of D are negative, and μ is given by any of the constant closure models (5.61), (5.62), (5.63), (5.64), (5.65), (5.66), or (5.67). Then, the nonlinear closure model

$$
H_{nl} = \mu_{nl}\tilde{f}(q^{pod})\, diag(d_{11}, \dots, d_{N_{pod}N_{pod}}) q^{pod}, \quad \mu_{nl} > 0 \tag{5.81}
$$

stabilizes the solutions of the ROM to the invariant set

$$
S = \left\{ q^{pod} \in \mathbb{R}^{N_{pod}} \text{ s.t. } \mu \frac{\lambda(D)_{max}\|q^{pod}\|}{\tilde{f}(q^{pod})} + \mu_{nl}\|q^{pod}\| Max(d_{11}, \dots, d_{N_{pod}N_{pod}}) \right.
$$
$$
\left. + 1 \geq 0 \right\}.
$$

Furthermore, if the closure models' amplitudes μ_e, μ_{nl} are tuned using the MES algorithm

$$
\dot{y}_1 = a_1 \sin\left(\omega_1 t + \frac{\pi}{2}\right) Q(\hat{\mu}_e, \hat{\mu}_{nl}),
$$
$$
\hat{\mu}_e = y_1 + a_1 \sin\left(\omega_1 t - \frac{\pi}{2}\right),
$$
$$
\dot{y}_2 = a_2 \sin\left(\omega_2 t + \frac{\pi}{2}\right) Q(\hat{\mu}_e, \hat{\mu}_{nl}), \tag{5.82}
$$
$$
\hat{\mu}_{nl} = y_2 + a_2 \sin\left(\omega_2 t - \frac{\pi}{2}\right),
$$

where $\omega_{\max} = \max(\omega_1, \omega_2) > \omega^*$, ω^* *large enough, and* Q *is given by Eq. (5.72), with* $\hat{\mu} = (\hat{\mu}_e, \hat{\mu}_{nl})$. *Then, under Assumptions 5.17 and 5.18, the norm of the vector of the distance with respect to the optimal values of* μ_e, μ_{nl}; $e_\mu = (\mu_e^* - \hat{\mu}_e(t), \mu_{nl}^* - \hat{\mu}_{nl}(t))$ *admits the following bound:*

$$\|e_\mu(t)\| \leq \frac{\xi_1}{\omega_{\max}} + \sqrt{a_1^2 + a_2^2}, \quad t \to \infty \qquad (5.83)$$

where $a_1, a_2 > 0, \xi_1 > 0$, *and the learning cost function approaches its optimal value within the following upper bound:*

$$|Q(\hat{\mu}_e, \hat{\mu}_{nl}) - Q(\mu_e^*, \mu_{nl}^*)| \leq \xi_2 \left(\frac{\xi_1}{\omega} + \sqrt{a_1^2 + a_2^2} \right), \quad t \to \infty \quad (5.84)$$

where $\xi_2 = \max_{(\mu_1, \mu_2)} \in \mathcal{V}(\mu^*) \|\frac{\partial Q}{\partial \mu}\|$.

Proof. First, we prove that the nonlinear closure model (5.81) stabilizes the ROM (5.80) to an invariant set. To do so, we use the following energy-like Lyapunov function:

$$V = \frac{1}{2} q^{\text{pod}^T} q^{\text{pod}}. \qquad (5.85)$$

We then evaluate the derivative of V along the solutions of Eqs. (5.80) and (5.81) (using Assumption 5.19) as

$$\dot{V} = q^{\text{pod}^T} (\tilde{F}(q^{\text{pod}}(t)) + \mu D q^{\text{pod}} + \mu_{nl} \tilde{f}(q^{\text{pod}}) \text{diag}(d_{11}, \ldots, d_{N_{\text{pod}} N_{\text{pod}}}) q^{\text{pod}}),$$

$$\leq \|q^{\text{pod}}\| \tilde{f}(q^{\text{pod}}) + \mu \|q^{\text{pod}}\|^2 \lambda(D)_{\max} + \mu_{nl} \tilde{f}(q^{\text{pod}}) \|q^{\text{pod}}\|^2 \text{Max}(d_{11}, \ldots, d_{N_{\text{pod}} N_{\text{pod}}}),$$

$$\leq \|q^{\text{pod}}\| \tilde{f}(q^{\text{pod}}) \left(1 + \mu \frac{\lambda(D)_{\max} \|q^{\text{pod}}\|}{\tilde{f}(q^{\text{pod}})} + \mu_{nl} \text{Max}(d_{11}, \ldots, d_{N_{\text{pod}} N_{\text{pod}}}) \|q^{\text{pod}}\| \right),$$

$$(5.86)$$

which shows the convergence to the invariant set

$$\mathcal{S} = \left\{ q^{\text{pod}} \in \mathbb{R}^{N_{\text{pod}}} \text{ s.t. } \mu \frac{\lambda(D)_{\max} \|q^{\text{pod}}\|}{\tilde{f}(q^{\text{pod}})} \right.$$
$$\left. + \mu_{nl} \|q^{\text{pod}}\| \text{Max}(d_{11}, \ldots, d_{N_{\text{pod}} N_{\text{pod}}}) + 1 \geq 0 \right\}.$$

The rest of the proof follows the same steps as in the proof of Lemma 5.6.

Remark 5.16. *The previous algorithms have been presented in the continuous time framework. However, their discreet counterpart, where the learning is done over a number of iterations* N, *each with a finite time support* t_f, *can be readily written, as in Section 5.5.1.2 for the case of PDEs open-loop parameters estimation.*

Remark 5.17. *In both Theorem 5.4 and Lemma 5.7, instead of choosing a scalar for the closure model amplitude μ_{nl}, we can choose to use a vector of amplitudes μ_{nl-i}, one for each POD mode $i \in \{1, \ldots, N_{pod}\}$; we will call the associated closure models in this case "modal" closure models. The generalization of the convergence results of Theorem 5.4 and Lemma 5.7 to the case of modal closure models is straightforward because the same convergence results of MES algorithms will apply in this case as well.*

As an illustrative example, to show what we mean by the fact that the previous formulation is suitable in the case of bounded model uncertainties, we show how the term $\tilde{f}$ is defined in the case of some specific PDEs' structures.

(A) The case of the Navier-Stokes equations

We consider here the case of the Navier-Stokes equations. We assume the case of incompressible, viscous flow in a steady spacial domain, with time-independent spacial boundary conditions, that is, Neumann conditions, or Dirichlet conditions, and so on. Under these conditions, a standard Galerkin projection of the Navier-Stokes equations onto POD modes leads to a POD ROM system of the following form, see, Cordier et al. (2013):

$$\dot{q}^{\text{pod}} = \mu\, D\, q^{\text{pod}} + [Cq^{\text{pod}}]q^{\text{pod}} + [Pq^{\text{pod}}]q^{\text{pod}},$$

$$u(x, t) = u_0(x) + \sum_{i=1}^{i=N_{\text{pod}}} \phi(x)_i^{\text{pod}} q_i^{\text{pod}}(t), \tag{5.87}$$

where $\mu > 0$ is the viscosity number, that is, inverse of the Reynolds number, D the viscosity damping matrix with negative diagonal elements, C the convection effect three-dimensional tensor, and P the pressure effect three-dimensional tensor. We notice that this POD ROM has mainly a linear term and two quadratic terms. The term u represents the velocity field, written as the sum of its mean value u_0 (or a basic mode) and the expansion of N_{pod} spacial modes ϕ_i^{pod}, $i = 1, \ldots, N^{\text{pod}}$.

The POD ROM (5.87) is of the form of Eq. (5.68), with

$$\tilde{F} = [Cq^{\text{pod}}]q^{\text{pod}} + [Pq^{\text{pod}}]q^{\text{pod}}.$$

If we consider bounded parametric uncertainties on the coefficients of C or P, we can write that

$$\tilde{F} = [(C + \Delta C)q^{\text{pod}}]q^{\text{pod}} + [(P + \Delta P)q^{\text{pod}}]q^{\text{pod}},$$

where $\|C + \Delta C\|_F \leq c_{\max}$ and $\|P + \Delta P\|_F \leq p_{\max}$, which leads to the following upper bound for $\tilde{F}$:

$$\|\tilde{F}\| \leq c_{\max}\|q^{pod}\|^2 + p_{\max}\|q^{pod}\|^2.$$

In this case the nonlinear closure model (5.81) writes as

$$H_{nl} = \mu_{nl}(c_{\max}\|q^{pod}\|^2 + p_{\max}\|q^{pod}\|^2)\text{diag}(d_{11},\dots,d_{N_{pod}N_{pod}})q^{pod}, \quad \mu_{nl} > 0 \tag{5.88}$$

with the $d_{ii}, i = 1,\dots,N^{pod}$ being the diagonal elements of D.

Remark 5.18. *Obviously, in the previous case the two quadratic terms* $c_{\max}\|q^{pod}\|^2$, *and* $p_{\max}\|q^{pod}\|^2$ *can be combined as one quadratic term* $Max(c_{\max}, p_{\max})\|q^{pod}\|^2$, *but we choose to explicitly keep the two terms to clearly show the effect of the uncertainties of each matrix C, and P distinctively.*

(B) The case of the Burgers' equations

Simplified Burgers' equations represents the velocity field using the following PDE, see for example, San and Iliescu (2013):

$$\frac{\partial u}{\partial t} = \mu\frac{\partial^2 u}{\partial^2 x} - u\frac{\partial u}{\partial x}, \tag{5.89}$$

where u represents the velocity field and $x \in \mathbb{R}$ the one-dimensional space variable. Under Dirichlet boundary condition the POD ROM can be written as, see, San and Iliescu (2013)

$$\dot{q}^{pod} = B_1 + \mu B_2 + \mu D q^{pod} + \tilde{D}q^{pod} + [Cq^{pod}]q^{pod},$$

$$u_{ROM}(x,t) = u_0(x) + \sum_{i=1}^{i=N_{pod}} \phi(x)_i^{pod} q_i^{pod}(t), \tag{5.90}$$

where $\mu > 0$ is the viscosity number, that is, inverse of the Reynolds number, D the viscosity damping matrix with negative diagonal elements, C the convection effect tensor, and $\tilde{D}, B_1, B_2$ are constant matrices. Compared to the previous case this POD ROM has a constant term, two linear terms (among which one is directly proportional to μ), and a quadratic term. Similar to the previous example, the term u represents the velocity field written as the sum of its mean value u_0 and the expansion of the spacial modes $\phi_i^{pod}, i = 1,\dots,N^{pod}$.

In this case to write the ROM (5.90) in the form of Eq. (5.68) we define $\tilde{F}$ as

$$\tilde{F} = B_1 + \mu B_2 + \tilde{D}q^{pod} + [Cq^{pod}]q^{pod},$$

which can be upper bounded by

$$\|\tilde{F}\| \leq b_{1_{\max}} + \mu_{\max}b_{2_{\max}} + \tilde{d}_{\max}\|q^{pod}\| + c_{\max}\|q^{pod}\|^2,$$

where $\|B_1 + \Delta B_1\|_F \leq b_{1_{\max}}$, $\|B_2 + \Delta B_2\|_F \leq b_{2_{\max}}$, $\mu \leq \mu_{\max}$, $\|\tilde{D} + \Delta\tilde{D}\|_F \leq \tilde{d}_{\max}$, and $\|C + \Delta C\|_F \leq c_{\max}$.

This leads to the nonlinear closure model

$$H_{\mathrm{nl}} = \mu_{\mathrm{nl}}(b_{1_{\max}} + \mu_{\max}b_{2_{\max}} + \tilde{d}_{\max}\|q^{\mathrm{pod}}\| + c_{\max}\|q^{\mathrm{pod}}\|^2)$$
$$\mathrm{diag}(d_{11}, \ldots, d_{N_{\mathrm{pod}}N_{\mathrm{pod}}})q^{\mathrm{pod}}. \tag{5.91}$$

Remark 5.19. *We see in the previous expression of H_{nl} that we need an upper bound for the viscosity coefficient μ. If H_{nl} is used alone as the sole closure model stabilizing Eq. (5.80), that is, μ is constant and nominal, then it is straightforward to fix an upper bound for μ. In the case where μ is given by one of the constant linear closure models (5.61), (5.62), (5.63), (5.64), (5.65), (5.66), or (5.67), then the upper bound of the amplitude μ_e needs to be taken into account as well. This can easily be done by assuming an upper bound of the search space for μ_e by the extremum seeker.*

In the next section, we will illustrate the theory developed in the previous sections of this chapter on several practical examples. We will apply the open-loop and closed-loop identification algorithms to two mechatronics examples, namely the electromagnetic actuators and a two-link rigid manipulator. We will also test the PDEs' identification and stabilization proposed approaches to the case of the simplified and the coupled Burgers' PDE.

5.6 APPLICATION EXAMPLES

5.6.1 Electromagnetic Actuator

We consider the following nonlinear model for electromagnetic actuators, see, Wang et al. (2000) and Peterson and Stefanopoulou (2004):

$$m\frac{d^2x_a}{dt^2} = k(x_0 - x_a) - \eta\frac{dx_a}{dt} - \frac{ai^2}{2(b + x_a)^2},$$
$$u = Ri + \frac{a}{b + x_a}\frac{di}{dt} - \frac{ai}{(b + x_a)^2}\frac{dx_a}{dt}, \quad 0 \leq x_a \leq x_f, \tag{5.92}$$

where x_a represents the armature position physically constrained between the initial position of the armature 0, and the maximal position of the armature x_f, $\frac{dx_a}{dt}$ represents the armature velocity, m is the armature mass, k the spring constant, x_0 the initial spring length, η the damping coefficient (assumed to be constant), $\frac{ai^2}{2(b+x_a)^2}$ represents the electromagnetic force (EMF) generated by the coil, a, b are two constant parameters of the coil, R the resistance of the coil, $L = \frac{a}{b+x_a}$ the coil inductance, $\frac{ai}{(b+x_a)^2}\frac{dx_a}{dt}$ represents the back EMF. Finally, i denotes the coil current, $\frac{di}{dt}$ its time derivative, and u represents the control voltage applied to the coil. In this model we do not consider the saturation region of the flux linkage in the magnetic field generated by the coil because we assume a current and armature motion ranges within the linear region of the flux.

We will use the results of open-loop identification presented in Section 5.3. We only present here simulation results, but in a real test-bed setting we would run in parallel the open-loop electromagnet driven by a bounded voltage over a finite time interval $[0, t_f]$, and then use measurements of the systems and measurements of the model to drive the MES-based identification algorithm of Lemma 5.1. The electromagnet system is clearly Lagrange stable, in the sense that a bounded voltage will lead to bounded coil current, armature positions, and armature velocity. In this case an open-loop identification is feasible. We report several tests where we target the identification of different parameters.

Test 1: We first consider the case where we want to identify the damping coefficient η. The identification is performed using the ES algorithm

$$\hat{\eta}(t) = \eta_{\text{nom}} + \Delta\eta(t),$$
$$\Delta\eta(t) = \delta\hat{\eta}((I - 1)t_f), (I - 1)t_f \leq t < It_f, \quad I = 1, 2, \ldots$$
$$\dot{z} = a\sin\left(\omega t + \frac{\pi}{2}\right) Q(\hat{\eta}), \quad (5.93)$$
$$\delta\hat{\eta} = z + a\sin\left(\omega t - \frac{\pi}{2}\right),$$

where η_{nom} is a given nominal value (the best known value beforehand), and I is the number of learning iterations. We assume that the actual value of the coefficient is $\eta = 7.5$ kg/s and that the assumed known value is $\eta_{\text{nom}} = 7$ kg/s. The remaining model parameters are set to $k = 158$ N/mm, $m = 0.27$ kg, $x_0 = 8$ mm, $R = 6\ \Omega$, $a = 14.96$ Nm$^2/A^2$, $b = 0.04$ mm. At each iteration we run the system in feedforward for a time interval length $t_f = 0.5$ s. The feedforward voltage is simply computed by imposing an armature smooth trajectory $y_d(t)$ and obtaining the associate current and feedforward voltage by direct inversion of the model (5.92). The MES parameters have been selected as follows: $a_1 = 6 \times 10^{-4}$, $a_2 = 10^{-2}$, $a_3 = 10^{-4}$, $\omega_1 = 5$ rad/s, $\omega_2 = 4$ rad/s, and $\omega_3 = 10$ rad/s. The learning cost function has been chosen as

$$Q(\delta\hat{\eta}) = \int_{(I-1)t_f}^{It_f} x_a^T(t)c_1 x_a(t)\,dt + \int_{(I-1)t_f}^{It_f} \dot{x}_a^T(t)c_2 \dot{x}_a(t)\,dt + \int_{(I-1)t_f}^{It_f} i^T(t)c_3 i(t)\,dt,$$
$$(5.94)$$

for each iteration $I = 1, 2, \ldots$, with $c_1 = c_2 = c_3 = 100$. We report the learning function profile over the iterations in Fig. 5.1A. We see a quick convergence of the cost function to its minimum. The associated parameter estimate is given in Fig. 5.1B, where we see that as expected the estimate converges to a small neighborhood of the actual value $\delta\eta = 0.5$.

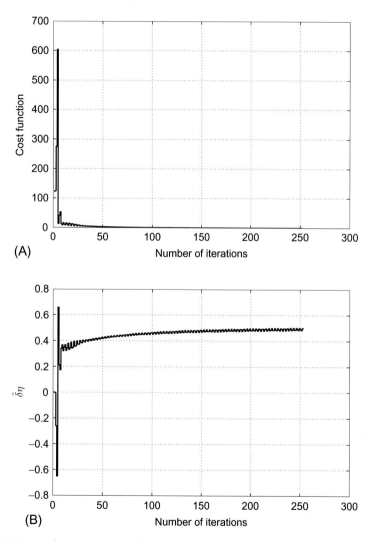

(A)

(B)

Fig. 5.1 Learned parameters and learning cost function: electromagnet identification Test 1. (A) Learning cost function versus number of iterations. (B) Learned parameter $\delta\hat{\eta}$ versus number of iterations.

Test 2: We test now a more challenging case, where the identified parameter appears in a nonlinear expression in the model. Indeed, we want to test the performance of the MES-based identification algorithms on the estimation of the parameter b in Eq. (5.92). We assume that the actual value is $b = 4 \times 10^{-2}$ mm, whereas the assumed known value is $b_{\text{nom}} = 3 \times 10^{-2}$ mm. The rest of the model's parameters are similar to the one in Test 1. The parameter b is then estimated using the ES algorithm

$$
\begin{aligned}
\hat{b}(t) &= b_{\text{nom}} + \Delta b(t), \\
\Delta b(t) &= \delta \hat{b}((I-1)t_f), \quad (I-1)t_f \leq t < It_f, \quad I = 1, 2, \ldots \\
\dot{z} &= a \sin\left(\omega t + \frac{\pi}{2}\right) Q(\hat{b}), \\
\delta \hat{b} &= z + a \sin\left(\omega t - \frac{\pi}{2}\right),
\end{aligned}
\tag{5.95}
$$

where $a = 10^{-5}$ and $\omega = 10$ rad/s. The learning cost function is the same as in Test 1. The convergence of the cost function as well as the learned value of δb are shown in Fig. 5.2.

Test 3: In this final test we want to evaluate the performance of the identification algorithm when dealing with multiple parameters' identification. We consider that the three parameters k, η, b are unknown. We assume that their true values are $k = 158$ N/mm, $\eta = 7.53$ kg/s, $b = 4 \times 10^{-2}$ mm, whereas their best known nominal values are $k = 153$ N/mm, $\eta = 7.33$ kg/s, and $b = 3 \times 10^{-2}$ mm. The parameters are estimated using the MES algorithm

$$
\begin{aligned}
\hat{\eta}(t) &= \eta_{\text{nom}} + \Delta \eta(t), \\
\Delta \eta(t) &= \delta \hat{\eta}((I-1)t_f), \\
\dot{z}_1 &= a_1 \sin\left(\omega_1 t + \frac{\pi}{2}\right) Q(\hat{\eta}, \hat{k}, \hat{b}), \\
\delta \hat{\eta} &= z_1 + a_1 \sin\left(\omega_1 t - \frac{\pi}{2}\right), \\
\hat{k}(t) &= k_{\text{nom}} + \Delta k(t), \\
\Delta k(t) &= \delta \hat{k}((I-1)t_f), \\
\dot{z}_2 &= a_2 \sin\left(\omega_2 t + \frac{\pi}{2}\right) Q(\hat{\eta}, \hat{k}, \hat{b}), \\
\delta \hat{k} &= z_2 + a_2 \sin\left(\omega_2 t - \frac{\pi}{2}\right), \\
\hat{b}(t) &= b_{\text{nom}} + \Delta b(t), \\
\Delta b(t) &= \delta \hat{b}((I-1)t_f), \\
\dot{z}_3 &= a_3 \sin\left(\omega_3 t + \frac{\pi}{2}\right) Q(\hat{\eta}, \hat{k}, \hat{b}), \\
\delta \hat{b} &= z_3 + a_3 \sin\left(\omega_3 t - \frac{\pi}{2}\right),
\end{aligned}
\tag{5.96}
$$

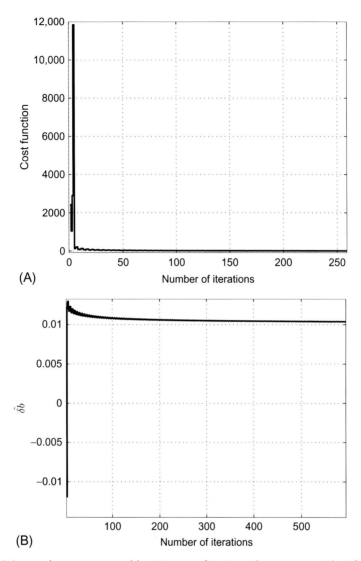

Fig. 5.2 Learned parameters and learning cost function: electromagnet identification Test 2. (A) Learning cost function versus number of iterations. (B) Learned parameter $\delta\hat{b}$ versus number of iterations.

for $(I - 1)t_f \leq t < It_f, I = 1, 2, \ldots,$ with $a_1 = 1.5 \times 10^{-5}, \omega_1 = 9$ rad/s, $a_2 = 15 \times 10^{-5}, \omega_2 = 8$ rad/s, $a_3 = 2 \times 10^{-6},$ and $\omega_3 = 10$ rad/s. The learning cost function is chosen as in the previous tests. The convergence of the cost function and the identified parameters is shown in Fig. 5.3. One point to underline here is that compared to the single parameter cases, the convergence takes relatively longer in the multiparametric case. This can be

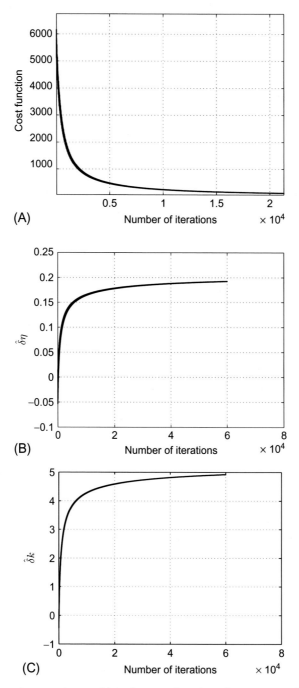

Fig. 5.3 Learned parameters and learning cost function: electromagnet identification Test 3. (A) Learning cost function versus number of iterations. (B) Learned parameter $\delta\hat{\eta}$ versus number of iterations. (C) Learned parameter $\delta\hat{k}$ versus number of iterations.

(Continued)

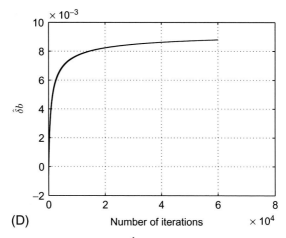

Fig. 5.3, Cont'd (D) Learned parameter $\delta \hat{b}$ versus number of iterations.

improved by better tuning the parameters of the MES algorithm, that is, the a_is and ω_is, see, Tan et al. (2008). Another approach is to use different MES algorithms which use, for example, extra filters to speed up the learning convergence and enlarge the basin of attraction, see, Noase et al. (2011).

5.6.2 Robot Manipulator With Two Rigid Arms

We consider a two-link robot manipulator with the following dynamics (see, e.g., Spong, 1992):

$$H(q)\ddot{q} + C(q, \dot{q})\dot{q} + G(q) = \tau, \tag{5.97}$$

where $q \triangleq [q_1, q_2]^T$ denotes the two joint angles and $\tau \triangleq [\tau_1, \tau_2]^T$ denotes the two joint torques. The matrix H is assumed to be nonsingular and is given by

$$H \triangleq \begin{bmatrix} H_{11} & H_{12} \\ H_{21} & H_{22} \end{bmatrix},$$

where

$$
\begin{aligned}
H_{11} &= m_1 \ell_{c_1}^2 + I_1 + m_2[\ell_1^2 + \ell_{c_2}^2 + 2\ell_1 \ell_{c_2} \cos(q_2)] + I_2, \\
H_{12} &= m_2 \ell_1 \ell_{c_2} \cos(q_2) + m_2 \ell_{c_2}^2 + I_2, \\
H_{21} &= H_{12}, \\
H_{22} &= m_2 \ell_{c_2}^2 + I_2.
\end{aligned}
\tag{5.98}
$$

The matrix $C(q, \dot{q})$ is given by

$$C(q, \dot{q}) \triangleq \left[\begin{array}{cc} -h\dot{q}_2 & -h\dot{q}_1 - h\dot{q}_2 \\ h\dot{q}_1 & 0 \end{array} \right],$$

where $h = m_2 \ell_1 \ell_{c_2} \sin(q_2)$. The vector $G = [G_1, G_2]^T$ is given by

$$\begin{aligned} G_1 &= m_1 \ell_{c_1} g \cos(q_1) + m_2 g [\ell_2 \cos(q_1 + q_2) + \ell_1 \cos(q_1)], \\ G_2 &= m_2 \ell_{c_2} g \cos(q_1 + q_2), \end{aligned} \tag{5.99}$$

where ℓ_1, ℓ_2 are the lengths of the first and second link, respectively; ℓ_{c_1}, ℓ_{c_2} are the distances between the rotation center and the center of mass of the first and second link, respectively. m_1, m_2 are the masses of the first and second link, respectively; I_1 is the moment of inertia of the first link and I_2 the moment of inertia of the second link; and g denotes the earth gravitational constant. We want to test the performance of the MES-based identification to estimate the values of the masses of the links. If we move the mass matrix H to the right-hand side of Eq. (5.97), we see that the joint angular accelerations are not linear functions of the parameters m_1, m_2. We use the following values for the robots' parameters: $I_1 = 0.92 \text{ kg m}^2$, $I_2 = 0.46 \text{ kg m}^2$, $\ell_1 = 1.1 \text{ m}$, $\ell_2 = 1.1 \text{ m}$, $\ell_{c_1} = 0.5 \text{ m}$, $\ell_{c_2} = 0.5 \text{ m}$, and $g = 9.8 \text{ m/s}^2$. We assume that the true values of the masses are $m_1 = 10.5 \text{ kg}$, $m_2 = 5.5 \text{ kg}$. The known nominal values are $m_1 = 9.5 \text{ kg}$, $m_2 = 5 \text{ kg}$. We identify the values of the masses using the MES algorithm

$$\begin{aligned} \hat{m}_1(t) &= m_{1\text{-nom}} + \Delta m_1(t), \\ \Delta m_1(t) &= \delta \hat{m}_1((I-1)t_f), \\ \dot{z}_1 &= a_1 \sin\left(\omega_1 t + \frac{\pi}{2}\right) Q(\hat{m}_1, \hat{m}_2), \\ \delta \hat{m}_1 &= z_1 + a_1 \sin\left(\omega_1 t - \frac{\pi}{2}\right), \\ \hat{m}_2(t) &= m_{2\text{-nom}} + \Delta m_2(t), \\ \Delta m_2(t) &= \delta \hat{m}_2((I-1)t_f), \\ \dot{z}_2 &= a_2 \sin\left(\omega_2 t + \frac{\pi}{2}\right) Q(\hat{m}_1, \hat{m}_2), \\ \delta \hat{m}_2 &= z_2 + a_2 \sin\left(\omega_2 t - \frac{\pi}{2}\right), \end{aligned} \tag{5.100}$$

with the parameters $a_1 = 3 \times 10^5$, $\omega_1 = 7 \text{ rad/s}$, $a_2 = 10^5$, and $\omega_2 = 5 \text{ rad/s}$. The profiles of the learning cost function and the estimated values of the masses' deltas are shown in Fig. 5.4.

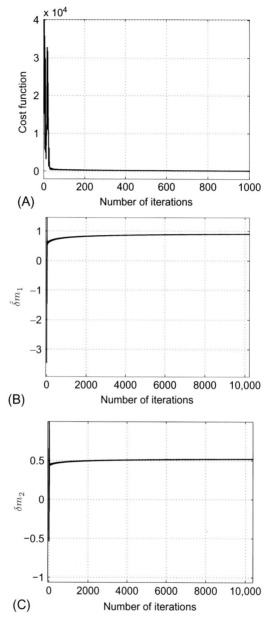

Fig. 5.4 Learned parameters and learning cost function: robot manipulator identification. (A) Learning cost function versus number of iterations. (B) Learned parameter $\delta\hat{m}_1$ versus number of iterations. (C) Learned parameter $\delta\hat{m}_2$ versus number of iterations.

5.6.3 The Coupled Burgers' PDE

We consider here the case of the coupled Burgers' equation, which is a generalization of the simplified equation given earlier in Eq. (5.89), and writes as, see, Kramer (2011)

$$\begin{cases} \frac{\partial \omega(t,x)}{\partial t} + \omega(t,x)\frac{\omega(t,x)}{\partial x} = \mu\frac{\partial^2 \omega(t,x)}{\partial x^2} - \kappa T(t,x) \\ \frac{\partial T(t,x)}{\partial t} + \omega(t,x)\frac{\partial T(t,x)}{\partial x} = c\frac{\partial^2 T(t,x)}{\partial x^2} + f(t,x), \end{cases} \tag{5.101}$$

where T represents the temperature and ω represents the velocity field, κ is the coefficient of the thermal expansion, c the heat diffusion coefficient, μ the viscosity coefficient (inverse of the Reynolds number R_e), x is the one-dimensional space variable $x \in [0,1]$, $t > 0$, and f is the external forcing term such that $f \in L^2((0,\infty), X), X = L^2([0,1])$. The previous equation is associated with the following boundary conditions:

$$\omega(t,0) = \delta_1, \quad \frac{\partial \omega(t,1)}{\partial x} = \delta_2, \tag{5.102}$$
$$T(t,0) = T_1, \quad T(t,1) = T_2,$$

where $\delta_1, \delta_2, T_1, T_2 \in \mathbb{R}_{\geq 0}$.

We consider here the following general initial conditions:

$$\omega(0,x) = \omega_0(x) \in L^2([0,1]), \tag{5.103}$$
$$T(0,x) = T_0(x) \in L^2([0,1]).$$

Following a Galerkin-type projection into POD basis functions, see for example, Kramer (2011), the coupled Burgers' equation is reduced to a POD ROM with the following structure:

$$\begin{pmatrix} \dot{q}_\omega^{pod} \\ \dot{q}_T^{pod} \end{pmatrix} = B_1 + \mu B_2 + \mu D q^{pod} + \tilde{D} q^{pod} + [C q^{pod}] q^{pod}$$

$$\omega_{ROM}(x,t) = \omega_0(x) + \sum_{i=1}^{i=N_{pod\omega}} \phi(x)_{\omega i}^{pod} q_{\omega i}^{pod}(t), \tag{5.104}$$

$$T_{ROM}(x,t) = T_0(x) + \sum_{i=1}^{i=N_{podT}} \phi(x)_{Ti}^{pod} q_{Ti}^{pod}(t),$$

where matrix B_1 is due to the projection of the forcing term f, matrix B_2 is due to the projection of the boundary conditions, D is due to the projection of the viscosity damping term $\mu\frac{\partial^2 \omega(t,x)}{\partial x^2}$, $\tilde{D}$ is due to the projection of the thermal coupling and the heat diffusion terms $-\kappa T(t,x)$, $c\frac{\partial^2 T(t,x)}{\partial x^2}$, and the tensor C is due to the projection of the gradient-based terms $\omega\frac{\omega(t,x)}{\partial x}$, $\omega\frac{\partial T(t,x)}{\partial x}$. The notations $\phi_{\omega i}^{pod}(x)$, $q_{\omega i}^{pod}(t)$ $(i = 1,\ldots,N_{pod\omega})$, $\phi_{Ti}^{pod}(x)$, $q_{Ti}^{pod}(t)$ $(i = 1,\ldots,N_{podT})$, stand for the space basis functions and the time projection coordinates, for the velocity and the temperature, respectively. $\omega_0(x)$, $T_0(x)$ represent the mean values (over time) of ω and T, respectively.

5.6.3.1 Burgers' Equation ES-Based Parameters' Estimation

To illustrate the ES-based parameters' estimation results presented in Section 5.5, we consider here the case of the Burgers' equation with several cases of uncertainties.

Test 1: First, we report the case with an uncertainty on the Reynolds number R_e. We consider the coupled Burgers' equation (5.101), with the parameters $R_e = 1000$, $\kappa = 1$, $c = 0.01$, the boundary conditions $\delta_1 = 0$, $\delta_2 = 5$, $T_1 = 0$, $T_2 = 0.1\sin(0.5\pi t)$, the initial conditions $\omega_0(x) = 2(x^2(0.5 - x)^2)$, $T_0(x) = 0.5\sin(\pi x)^5$, and a zero forcing term f. We assume a large uncertainty on R_e, and consider that its known value is $R_{e\text{-nom}} = 50$. We apply the estimation algorithm of Lemma 5.5. We estimate the value of R_e as follows:

$$\hat{R}_e(t) = R_{e\text{-nom}} + \delta R_e(t),$$
$$\delta R_e(t) = \delta\hat{R}_e((I - 1)t_f), \quad (I - 1)t_f \leq t < It_f, \qquad I = 1,2,\ldots \tag{5.105}$$

where I is the iteration number, $t_f = 50$ s the time horizon of one iteration, and $\hat{\delta}R_e$ is computed using the iterative ES algorithm

$$\dot{z} = a\sin\left(\omega t + \frac{\pi}{2}\right)Q(\hat{R}_e),$$
$$\delta\hat{R}_e = z_i + a\sin\left(\omega t - \frac{\pi}{2}\right). \tag{5.106}$$

We choose the learning cost function as

$$Q = Q_1\int_0^{t_f}\langle e_T, e_T\rangle dt + Q_2\int_0^{t_f}\langle e_\omega, e_\omega\rangle dt, \quad Q_1, Q_2 > 0 \tag{5.107}$$

with $e_T = T - T_{ROM}$, $e_\omega = \omega - \omega_{ROM}$ being the errors between the true model solution and the POD ROM solution for temperature and velocity, respectively. We applied the ES algorithm (5.106), (5.107), with $a = 0.0032$, $\omega = 10$ rad/s, $Q_1 = Q_2 = 1$.

We first show in Fig. 5.5 the plots of the true (obtained by solving the Burgers' PDE with finite elements method, with a grid of 100 elements in time and space[2]). We show in Fig. 5.6 the velocity and temperature profiles obtained by the nominal, that is, learning-free POD ROM with four POD modes for the velocity and four modes for the temperature, considering the incorrect parameter value $R_e = 50$. From Figs. 5.5 and 5.6, we can see that the temperature profile obtained by the nominal POD ROM is not too different from the true profile. However, the velocity

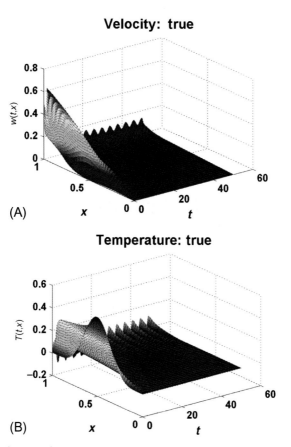

Fig. 5.5 True solutions of Eq. (5.101): Test 1. (A) True velocity profile. (B) True temperature profile.

[2]We thank here Dr. Boris Kramer, former intern at MERL, for sharing his codes to solve the Burgers' equation.

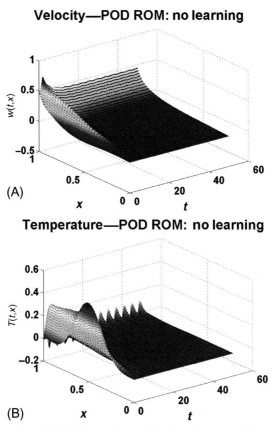

Fig. 5.6 Learning-free POD ROM solutions of Eq. (5.101): Test 1. (A) Learning-free POD ROM velocity profile. (B) Learning-free POD ROM temperature profile.

profiles are different, which is due to the fact that the uncertainty of R_e affects mainly the velocity part of the PDE. The error between the true solutions and the nominal POD ROM solutions are displayed in Fig. 5.7. Now, we show the ES-based learning of the uncertain parameter R_e. We first report in Fig. 5.8A the learning cost function over the learning iterations. We notice that with the chosen learning parameters a, ω the ES exhibits a big exploration step after the first iteration, which leads to a large cost function first. However, this large value of the cost function (due to the large exploration step) leads quickly to the neighborhood of the true value of R_e, as seen in Fig. 5.8B. The error between the POD ROM after learning and the true solutions are depicted in Fig. 5.9. By comparing Figs. 5.7 and 5.9, we can see that the error between the

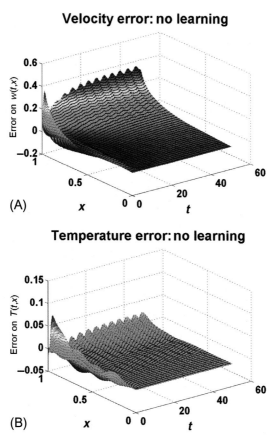

Fig. 5.7 Errors between the nominal POD ROM and the true solutions: Test 1. (A) Error between the true velocity and the learning-free POD ROM velocity profile. (B) Error between the true temperature and the learning-free POD ROM temperature profile.

POD ROM solutions and the true solutions have been largely reduced with the learning of the actual value of R_e, even though the convergence of $\hat{R}_e$ is only to a neighborhood of the true value of R_e, as stated in Lemma 5.5.

Test 2: Let us consider now the case of uncertainty on the coupling coefficient κ. We test the case where its actual value is $\kappa = 1$, whereas its assumed known value is $\kappa_{\mathrm{nom}} = 0.5$. The rest of the coefficients, number of POD modes, and boundary conditions remain similar to the first test. We show first the error between the learning-free POD ROM solutions and the true solutions in this case, reported in Fig. 5.10. We then estimate

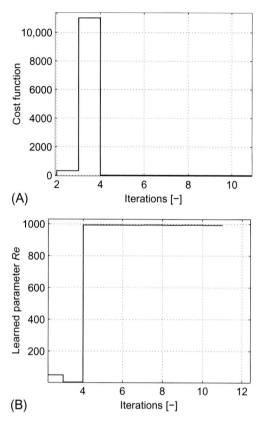

Fig. 5.8 Learned parameters and learning cost function: Test 1. (A) Learning cost function versus number of iterations. (B) Learned parameter $\hat{R}_e$ versus number of iterations.

the value of κ using the ES algorithm

$$\hat{\kappa}(t) = \kappa_{\text{nom}} + \delta\kappa(t),$$
$$\delta\kappa(t) = \delta\hat{\kappa}((I-1)t_f), \quad (I-1)t_f \le t < It_f, \quad I = 1, 2, \dots \tag{5.108}$$

where I is the iteration number, $t_f = 50$ s the time horizon of one iteration, and $\delta\hat{\kappa}$ is computed using the iterative ES algorithm

$$\dot{z} = a\sin\left(\omega t + \frac{\pi}{2}\right) Q(\hat{\kappa}),$$
$$\delta\hat{\kappa} = z_i + a\sin\left(\omega t - \frac{\pi}{2}\right), \tag{5.109}$$

with $a = 1.8 \times 10^{-4}, \omega = 5$ rad/s, and Q is given by Eq. (5.107), with $Q_1 = Q_2 = 1$. The learning cost function profile and the learned

Velocity error: with learning

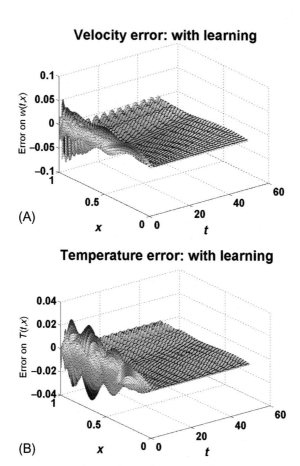

(A)

Temperature error: with learning

(B)

Fig. 5.9 Errors between the learning-based POD ROM and the true solutions: Test 1. (A) Error between the true velocity and the learning-based POD ROM velocity profile. (B) Error between the true temperature and the learning-based POD ROM temperature profile.

parameter $\hat{\kappa}$ are reported in Fig. 5.11. We see a quick convergence of the cost function to its minimum, and of the estimated parameter $\hat{\kappa}$ to a neighborhood of its value. We underline here that the upper bound of the estimation error can be tightened by using a more "sophisticated" ES algorithm, for example, ES with time-varying perturbation signal amplitude a, or with extra integration gains, for example, see Moase et al. (2009). The errors between the learning-based POD ROM and the true solutions are reported in Fig. 5.12, where the errors are shown to be smaller than in the learning-free case of Fig. 5.10.

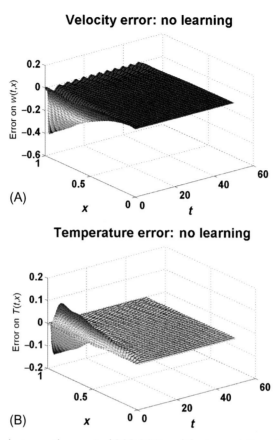

Fig. 5.10 Errors between the nominal POD ROM and the true solutions: Test 2. (A) Error between the true velocity and the learning-free POD ROM velocity profile. (B) Error between the true temperature and the learning-free POD ROM temperature profile.

Test 3: In the third test, we consider an uncertainty on the heat diffusion coefficient c, which influences directly the temperature equation in the coupled Burgers' equation. We consider the case where the true value is $c = 0.01$, whereas the assumed known value is $c_{\text{nom}} = 0.1$. The rest of the coefficients and number of POD modes remains unchanged. In this case the estimation of c is done with the ES algorithm

$$\hat{c}(t) = c_{\text{nom}} + \delta c(t),$$
$$\delta c(t) = \delta \hat{c}((I-1)t_f), \quad (I-1)t_f \leq t < It_f, \qquad I = 1, 2, \ldots \tag{5.110}$$

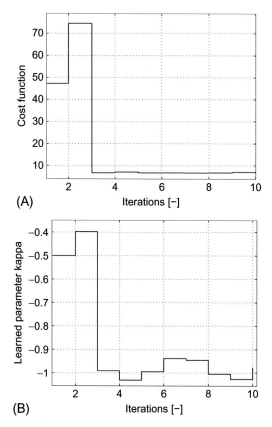

Fig. 5.11 Learned parameters and learning cost function: Test 2. (A) Learning cost function versus number of iterations. (B) Learned parameter $\hat{\kappa}$ versus number of iterations.

where I is the iteration number, $t_f = 50$ s the time horizon of one iteration, and $\delta\hat{c}$ is computed using the iterative ES algorithm

$$\dot{z} = a \sin\left(\omega t + \frac{\pi}{2}\right) Q(\hat{c}),$$

$$\delta\hat{c} = z_i + a \sin\left(\omega t - \frac{\pi}{2}\right),$$

(5.111)

with $a = 1.2 \times 10^{-5}, \omega = 10$ rad/s, and Q is given by Eq. (5.107), with $Q_1 = Q_2 = 1$. As in the previous test, we start by reporting the errors between the nominal POD ROM solutions and the true solutions in Fig. 5.13. The cost function profile as well as the estimation of c are reported in Fig. 5.14, where we see the convergence of $\hat{c}$ to the actual value of c, within an expected estimation error bound. The learning-based POD

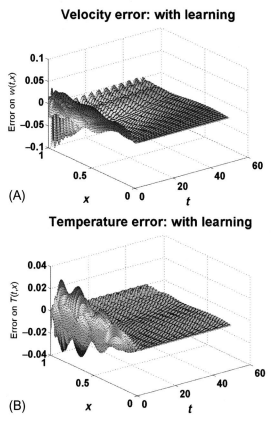

Fig. 5.12 Errors between the learning-based POD ROM and the true solutions: Test 2. (A) Error between the true velocity and the learning-based POD ROM velocity profile. (B) Error between the true temperature and the learning-based POD ROM temperature profile.

ROM gives a better prediction of the true solutions indicated in Fig. 5.15, where we show the errors between the learning-based ROM solutions and the true solutions.

Let us now consider the more challenging problem of Burgers' equation stabilization, using the results proposed in Section 5.5.2.

5.6.3.2 Burgers' Equation ES-Based POD ROM Stabilization

Test 1: First, we report the case related to Lemma 5.7, where we test the auto-tuning results of MES on the combination of a linear constant viscosity

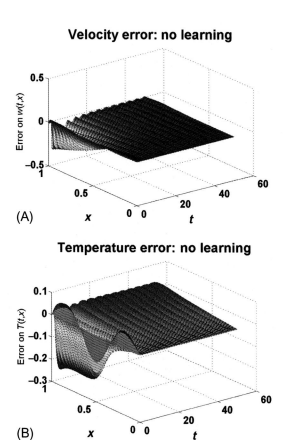

Velocity error: no learning

Temperature error: no learning

Fig. 5.13 Errors between the nominal POD ROM and the true solutions: Test 3. (A) Error between the true velocity and the learning-free POD ROM velocity profile. (B) Error between the true temperature and the learning-free POD ROM temperature profile.

closure model and a nonlinear closure model. We consider the coupled Burgers' equation (5.101), with the parameters $R_e = 1000, \kappa = 5 \times 10^{-4}, c = 0.01$, the trivial boundary conditions $\delta_1 = \delta_2 = 0, T_1 = T_2 = 0$, a simulation time length $t_f = 1$ s, and a zero forcing term f, we use 10 POD modes for both the temperature and the velocity variables. For the choice of the initial conditions we follow the conditions proposed in San and Iliescu (2013), where the simplified Burgers' equation has been used in the context of POD ROM stabilization. Indeed, in San and Iliescu (2013) the authors propose two types of initial conditions for the velocity variable, which led to instability of the nominal POD ROM, that is, the basic Galerkin POD ROM (POD ROM-G) without any closure model. Thus we follow

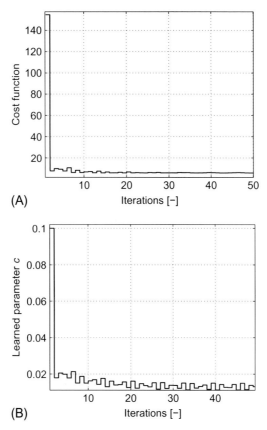

Fig. 5.14 Learned parameters and learning cost function: Test 3. (A) Learning cost function versus number of iterations. (B) Learned parameter $\hat{c}$ versus number of iterations.

San and Iliescu (2013) in the choice of the initial conditions, which make the coupled Burgers' equation model reduction a challenging problem. We choose the following initial conditions:

$$\omega(x,0) = \begin{cases} 1, & \text{if } x \in [0,0.5], \\ 0, & \text{if } x \in]0.5,1], \end{cases} \tag{5.112}$$

$$T(x,0) = \begin{cases} 1, & \text{if } x \in [0,0.5], \\ 0, & \text{if } x \in]0.5,1]. \end{cases} \tag{5.113}$$

We apply Lemma 5.7, with the Heisenberg linear closure model given by Eq. (5.61). The two closure model amplitudes μ_e and μ_{nl} are tuned using the MES algorithm given by Eq. (5.82), with the values $a_1 =$

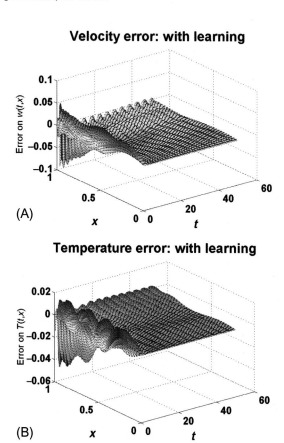

Fig. 5.15 Errors between the learning-based POD ROM and the true solutions: Test 3.
(A) Error between the true velocity and the learning-based POD ROM velocity profile.
(B) Error between the true temperature and the learning-based POD ROM temperature
profile.

$6 \times 10^{-6}, \omega_1 = 10$ rad/s, $a_2 = 6 \times 10^{-6}, \omega_2 = 15$ rad/s. The learning
cost function has been selected as in Eq. (5.107), with $Q_1 = Q_2 = 1$.

We first show for comparison purposes the true solutions, obtained
by solving the PDE (5.101) with an FEM algorithm with a grid of 100
elements for time and 100 elements for the space domain. The true
solutions are reported in Fig. 5.16. We then report in Fig. 5.17 the solutions
of the POD ROM-G (without learning). We can see clearly on the figures
that the POD ROM-G solutions are not as smooth as the true solutions;
particularly the velocity profile is very irregular, and the goal of the closure
model tuning is to smooth both profiles, as much as the closure model
allows. For a clearer evaluation, we also report the errors between the true
solutions and the POD ROM-G solutions; in Fig. 5.18. We then apply

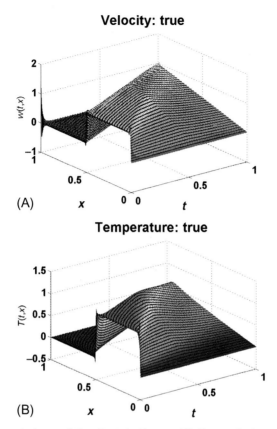

Fig. 5.16 True solutions of Eq. (5.101): Test 1. (A) True velocity profile. (B) True temperature profile.

the MES algorithm and show the profile of the learning cost function over the learning iterations in Fig. 5.19A. We can see a quick decrease of the cost function within the first 20 iterations. This means that the MES manages to improve the overall solutions of the POD ROM very fast. The associated profiles for the two closure models' amplitudes values $\hat{\mu}_e$ and $\hat{\mu}_{nl}$ are reported in Fig. 5.19B and C. We can see that even though the cost function value drops quickly, the MES algorithm continues to fine-tune the values of the parameters $\hat{\mu}_e$, $\hat{\mu}_{nl}$ over the iterations, and they eventually reach optimal values of $\hat{\mu}_e \simeq 0.3$, and $\hat{\mu}_{nl} \simeq 0.76$. We also show the effect of the learning on the POD ROM solutions in Figs. 5.20 and 5.21, which by comparison with Figs. 5.17 and 5.18 show a clear improvement of the POD ROM solutions with the MES tuning of the closure models.

Velocity—POD ROM: no learning

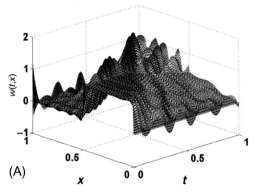

(A)

Temperature—POD ROM: no learning

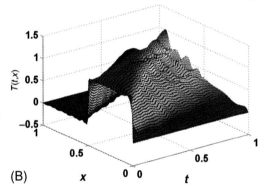

(B)

Fig. 5.17 Learning-free POD ROM solutions of Eq. (5.101): stabilization Test 1. (A) Learning-free POD ROM velocity profile. (B) Learning-free POD ROM temperature profile.

Test 2: We now evaluate the effect of the nonlinear closure model introduced in Theorem 5.4. In the Burgers' equation case the closure model (5.81) writes as Eq. (5.91). We use exactly the same coefficients for the Burgers' equation as in Test 1. We tune the two amplitudes μ_e, μ_{nl} of the closure modes using the MES algorithm given by Eq. (5.82), with the values $a_1 = 8 \times 10^{-6}, \omega_1 = 10$ rad/s, $a_2 = 8 \times 10^{-6}, \omega_2 = 15$ rad/s. The learning cost function has been selected as in Eq. (5.107) with $Q_1 = Q_2 = 1$. We recall that, for comparison purposes, the true profiles of the PDE solutions and the profiles of the POD ROM-G solutions are reported in Figs. 5.5 and 5.6. Next, the learning cost function value as a function of the learning iterations are given in Fig. 5.22A. The associated profiles of the learned

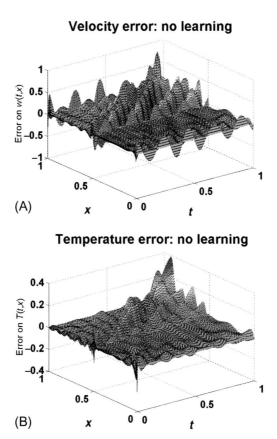

Fig. 5.18 Errors between the nominal POD ROM and the true solutions: stabilization Test 1. (A) Error between the true velocity and the learning-free POD ROM velocity profile. (B) Error between the true temperature and the learning-free POD ROM temperature profile.

parameters $\hat{\mu}_e$ and $\hat{\mu}_{nl}$ are shown in Fig. 5.22B and C. We see a clear convergence of the learning cost function and the learned parameters. The solutions of the POD ROM with MES auto-tuning are shown in Fig. 5.23, and the corresponding errors between the true solutions and the learning-based POD ROM are reported in Fig. 5.24. We see a clear improvement of the errors compared to the POD ROM-G. We can also see a slight improvement of the errors (especially the velocity error) compared to the learning-based POD ROM obtained when using the closure models of Lemma 5.7, reported in Test 1 (the velocity error in Fig. 5.24A is in the yellow spectrum whereas the one in Fig. 5.21A is in the green amplitude spectrum).

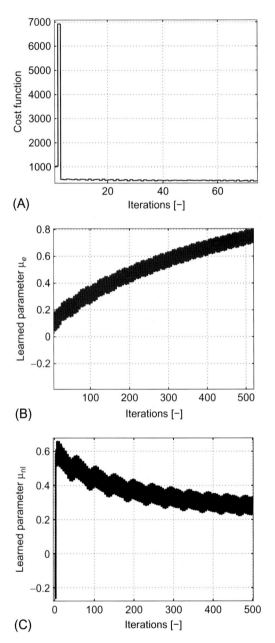

Fig. 5.19 Learned parameters and learning cost function: stabilization Test 1. (A) Learning cost function versus number of iterations. (B) Learned parameter $\hat{\mu}_e$ versus number of iterations. (C) Learned parameter $\hat{\mu}_{nl}$ versus number of iterations.

Velocity—POD ROM: with learning

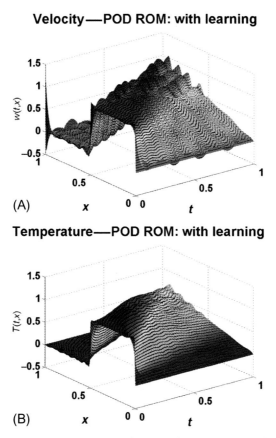

(A)

Temperature—POD ROM: with learning

(B)

Fig. 5.20 Learning-based POD ROM solutions of Eq. (5.101): stabilization Test 1. (A) Learning-based POD ROM velocity profile. (B) Learning-based POD ROM temperature profile.

Test 3: We want now to see if the use of a modal closure model, as introduced in Remark 5.17, will improve the overall performance of the MES-based ROM stabilization algorithms. We will test the same case as in Test 2, but here we use the algorithm of Theorem 5.4 with a modal closure model. In this case we have one linear closure model amplitude μ_e, one nonlinear closure model amplitude for the velocity POD modes μ_{nl-1}, and one nonlinear closure model amplitude for the temperature POD modes μ_{nl-2}. We choose not to use a full modal closure model, that is, a different amplitude for each mode, because we have 20 modes that will lead to 21 parameters to tune, which might lead to a slow convergence of the MES tuning algorithm. The closure models' amplitudes are then tuned using the MES

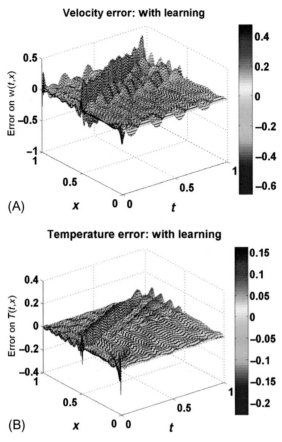

Fig. 5.21 Errors between the learning-based POD ROM and the true solutions: stabilization Test 1. (A) Error between the true velocity and the learning-based POD ROM velocity profile. (B) Error between the true temperature and the learning-based POD ROM temperature profile.

$$\dot{y}_1 = a_1 \, \sin\left(\omega_1 t + \frac{\pi}{2}\right) Q(\hat{\mu}_e, \hat{\mu}_{nl-1}, \hat{\mu}_{nl-2}),$$

$$\hat{\mu}_e = y_1 + a_1 \, \sin\left(\omega_1 t - \frac{\pi}{2}\right),$$

$$\dot{y}_2 = a_2 \, \sin\left(\omega_2 t + \frac{\pi}{2}\right) Q(\hat{\mu}_e, \hat{\mu}_{nl-1}, \hat{\mu}_{nl-2}),$$

$$\hat{\mu}_{nl-1} = y_2 + a_2 \, \sin\left(\omega_2 t - \frac{\pi}{2}\right), \qquad (5.114)$$

$$\dot{y}_3 = a_3 \, \sin\left(\omega_3 t + \frac{\pi}{2}\right) Q(\hat{\mu}_e, \hat{\mu}_{nl-1}, \hat{\mu}_{nl-2}),$$

$$\hat{\mu}_{nl-2} = y_3 + a_3 \, \sin\left(\omega_3 t - \frac{\pi}{2}\right).$$

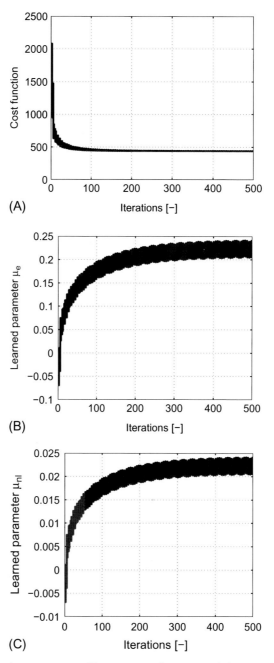

Fig. 5.22 Learned parameters and learning cost function: stabilization Test 2. (A) Learning cost function versus number of iterations. (B) Learned parameter $\hat{\mu}_e$ versus number of iterations. (C) Learned parameter $\hat{\mu}_{nl}$ versus number of iterations.

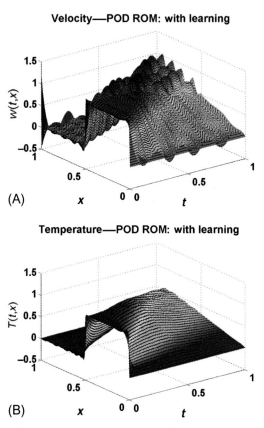

Fig. 5.23 Learning-based POD ROM solutions of Eq. (5.101): stabilization Test 2. (A) Learning-based POD ROM velocity profile. (B) Learning-based POD ROM temperature profile.

We implement Eq. (5.114) with the following coefficients: $a_1 = 8 \times 10^{-6}, \omega_1 = 10$ rad/s, $a_2 = 8 \times 10^{-6}, \omega_2 = 15$ rad/s, $a_3 = 8 \times 10^{-5}, \omega_3 = 12$ rad/s. A similar learning cost function to the previous tests is used here. We first report in Fig. 5.25 the learning cost function profile, as well as the learned parameters. Next, we report in Fig. 5.26 the POD ROM solutions. We can see some improvement of the velocity and temperature profiles. This improvement is even clearer if we look at the errors, plots in Fig. 5.27.

Test 4: Because we used some model upper bounds to write the closure model in Theorem 5.4, we wanted to test the robustness of the nonlinear closure model (5.81) compared with the nonlinear closure model used in Lemma 5.7. To do so, we conduct here the following test. We run both

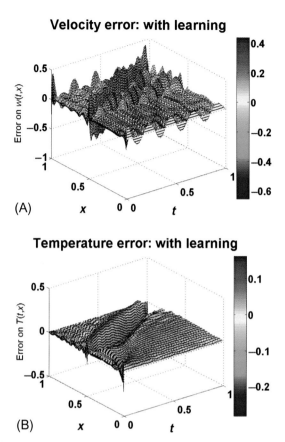

Fig. 5.24 Errors between the learning-based POD ROM and the true solutions: stabilization Test 2. (A) Error between the true velocity and the learning-based POD ROM velocity profile. (B) Error between the true temperature and the learning-based POD ROM temperature profile.

algorithms of Theorem 5.4 and Lemma 5.7 for some given values of the parameters R_e, κ, c of the Burgers' model (5.101). Then, we use the optimal closure models obtained by each algorithm and test them on a POD ROM with a different set of parameters R_e, κ, c. First, we assume that the known values of the Burgers' model parameters are $R_e = 1000, \kappa = -5 \times 10^{-4}$, $c = 0.01$, and then compute the POD ROM and use Theorem 5.4 to compute its optimal nonlinear closure model (5.81). After the convergence of the learning algorithm, the value of the learning cost function is $Q_{nominal} = 440.22$. Now, we test the same optimal closure model (5.81), but with a different POD ROM, corresponding to the parameters

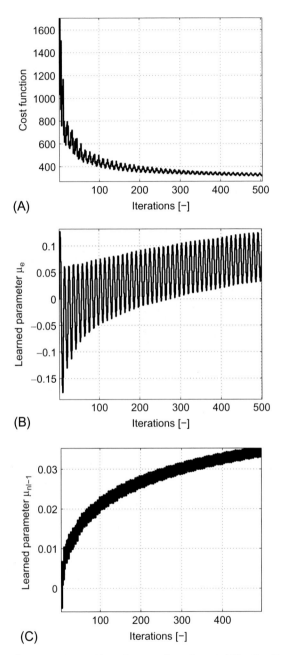

Fig. 5.25 Learned parameters and learning cost function: stabilization Test 3. (A) Learning cost function versus number of iterations. (B) Learned parameter $\hat{\mu}_e$ versus number of iterations. (C) Learned parameter $\hat{\mu}_{nl-1}$ versus number of iterations.

(Continued)

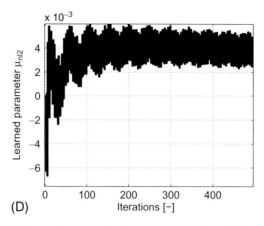

Fig. 5.25, Cont'd (D) Learned parameter $\hat{\mu}_{nl-2}$ versus number of iterations.

$R_e = 1800, \kappa = -5 \times 10^{-4}, c = 0.05$. The new cost function value is $Q_{\text{uncertain}} = 1352.3$. It is clear from the value of the cost function that the performance of the optimal closure model deteriorates. The corresponding plots of the velocity and temperature errors are given in Fig. 5.28. Next, we perform a similar test on the closure model of Lemma 5.7. The value of the learning cost function after convergence of learning algorithm, that is, in the nominal case, is $Q = 318.47$. Then, we test the same optimal closure model (5.69), but with a different POD ROM, corresponding to the parameters $R_e = 1800, \kappa = -5 \times 10^{-4}, c = 0.05$. The new cost function value is $Q_{\text{uncertain}} = 2026.7$. The degradation of the closure model performance is worse than in the case of the closure model (5.81). Indeed, we report in Fig. 5.29 the velocity and temperature errors in this case, and if we compare them to Fig. 5.28, we can see that the temperature error is worse in this case. The velocity errors profiles are only slightly different. We cannot conclude that the nonlinear closure model of Theorem 5.4 is more robust than the closure model of Lemma 5.7 because both of them perform well in keeping the ROM bounded, but the closure model of Theorem 5.4 seems to show more robustness to parametric uncertainties due to the fact that it has been explicitly computed based on some parametric uncertainties' upper bounds.

The reader can be wondering at this point why we are interested in robust closure models in the first place. We can agree with such a "philosophical" question in the sense that the problem of finding stable ROMs is intrinsically a mathematical problem, which means that we should

Velocity—POD ROM: with learning

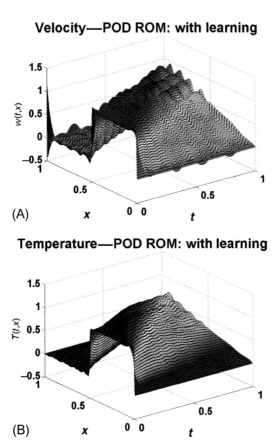

(A)

Temperature—POD ROM: with learning

(B)

Fig. 5.26 Learning-based POD ROM solutions of Eq. (5.101): stabilization Test 3. (A) Learning-based POD ROM velocity profile. (B) Learning-based POD ROM temperature profile.

have in our procession a perfectly known PDE model and try to find the best closure model for its ROM. However, we think that beyond the mathematical problem, where the learning algorithms presented in Section 5.5.2.2 are used offline, one could use these auto-tuning algorithms online as well. Indeed, we can start by computing a (POD) ROM offline and use the algorithms proposed here to auto-tune its closure models offline, but then we can implement the obtained "optimal" (POD) ROM in an online setting, for instance for control and estimation, and then continue to fine-tune its closure models online using the same MES-based algorithms. In this case, if the values of the parameters used in the original PDE model, and thus in the offline tuning of the closure models, are

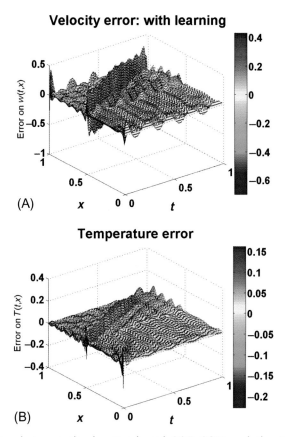

Fig. 5.27 Errors between the learning-based POD ROM and the true solutions: stabilization Test 3. (A) Error between the true velocity and the learning-based POD ROM velocity profile. (B) Error between the true temperature and the learning-based POD ROM temperature profile.

different from the actual values of the parameters of the real system, then the robustness of the closure model to such uncertainties will allow the ROM to stay stable while the MES-based algorithm will retune the closure models online to better fit them to the real system.

5.7 CONCLUSION AND OPEN PROBLEMS

In this chapter we have studied the problem of nonlinear systems identification. We have considered the case of open–loop Lagrange stable systems and have shown how ES can be used to estimate parameters of the system.

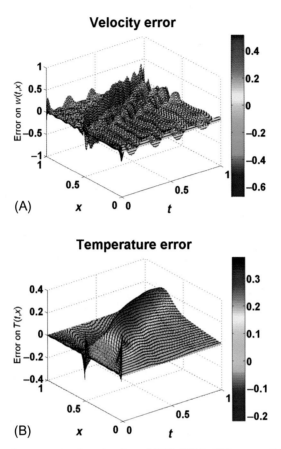

Fig. 5.28 Errors between the learning-based POD ROM of Theorem 5.4 and the true solutions: robustness test—stabilization Test 4. (A) Error between the true velocity and the learning-based POD ROM velocity profile. (B) Error between the true temperature and the learning-based POD ROM temperature profile.

We also considered the case of open-loop unstable systems, and shown that under stabilizing closed-loop feedback the parameters of the system can be estimated using ES. One of the most important features of the proposed identification algorithms is the fact that they can deal with parameters appearing nonlinearly in the model, which is well known to be a challenging case in identification. We also considered the problem of parametric identification for PDE systems, where we used ES-based methods to identify parameters of PDEs based on real-time measurements and ROMs. Finally, we have proposed some methods to solve the problem

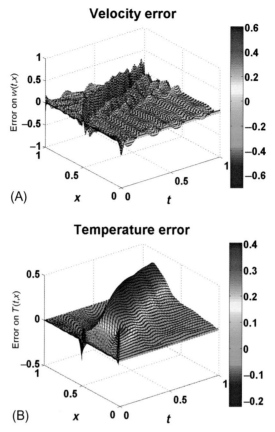

Fig. 5.29 Errors between the learning-based POD ROM of Lemma 5.7 and the true solutions: robustness test—stabilization Test 5. (A) Error between the true velocity and the learning-based POD ROM velocity profile. (B) Error between the true temperature and the learning-based POD ROM temperature profile.

of stable model reduction for PDEs. We have proposed ES-based auto-tuning algorithms to stabilize the ROMs. We have shown the performance of the proposed methods on several examples. The methods proposed here have shown great potential; however, we think that the performance of these learning-based identification and ROM stabilization methods could be further improved by using other types of learning approaches. For instance, reinforcement learning algorithms could be used instead of ES in these types of learning-based identification algorithms, for example to deal with the case of noisy measurements with high level of noise.

REFERENCES

Ahuja, S., Rowley, C., 2010. Feedback control of unstable steady states of flow past a flat plate using reduced-order estimators. J. Fluid Mech. 645, 447–478.

Andrieu, C., Doucet, A., Singh, S.S., Tadic, V.B., 2004. Particle methods for change detection, system identification and control. Proc. IEEE 92 (3), 423–438.

Astrom, K., Eykhoff, P., 1971. System identification—a survey. Automatica 7, 123–162.

Aubry, N., Holmes, P., Lumley, J.L., Stone, E., 1988. The dynamics of coherent structures in the wall region of a turbulent boundary layer. J. Fluid Mech. 192, 115–173.

Balajewicz, M., 2013. Lyapunov stable Galerkin models of post-transient incompressible flows, Tech. rep., arXiv.org/physics /arXiv:1312.0284.

Balajewicz, M., Dowell, E., Noack, B., 2013. Low-dimensional modelling of high-Reynolds-number shear flows incorporating constraints from the Navier-Stokes equation. J. Fluid Mech. 729 (1), 285–308.

Barbagallo, A., Sipp, D., Schmid, P.J., 2009. Closed-loop control of an open cavity flow using reduced-order models. J. Fluid Mech. 641, 1–50.

Barkana, I., 2014. Simple adaptive control—a stable direct model reference adaptive control methodology—brief survey. Int. J. Adapt. Control Signal Process. 28, 567–603.

Bendat, J., 1990. Nonlinear System Analysis and Identification from Random Data. Wiley Interscience, New York.

Bergmann, M., Bruneau, C., Iollo, A., 2009. Enablers for robust POD models. J. Comput. Phys. 228 (2), 516–538.

Calli, B., Caarls, W., Jonker, P., Wisse, M., 2012. Comparison of extremum seeking control algorithms for robotic applications. In: International Conference on Intelligent Robots and Systems (IROS). IEEE/RSJ, pp. 3195–3202.

Chollet, J.P., 1984. Two-point closure used for a sub-grid scale model in large eddy simulations. In: Turbulent Shear Flows 4. Springer, Berlin, pp. 62–72.

Colombo, R.M., Rosini, M.D., 2005. Pedestrian flows and non-classical shocks. Math. Methods Appl. Sci. 28 (13), 1553–1567.

Cordier, L., Noack, B., Tissot, G., Lehnasch, G., Delville, J., Balajewicz, M., Daviller, G., Niven, R.K., 2013. Identification strategies for model-based control. Exp. Fluids 54 (1580), 1–21.

Doucet, A., Tadic, V., 2003. Parameter estimation in general state-space models using particle methods. Ann. Inst. Stat. Math. 55, 409–422.

Fletcher, C.A.J., 1983. The group finite element formulation. Comput. Methods Appl. Mech. Eng. 37, 225–244.

Gevers, M., 2006. A personal view of the development of system identification a 30-year journey through an exciting field. IEEE Control Syst. Mag. 93–105.

Glad, S., Ljung, L., 1990a. Model structure identifiability and persistence of excitation. In: IEEE, Conference on Decision and Control, Honolulu, Hawaii, pp. 3236–3240.

Glad, S., Ljung, L., 1990b. Parametrization of nonlinear model structures as linear regressions. In: 11th IFAC World Congress, Tallinn, Estonia, pp. 67–71.

Golub, G., Van Loan, C., 1996. Matrix Computations, third ed. The Johns Hopkins University Press, Baltimore, MD.

Guay, M., Hariharan, N., 2008. Airflow velocity estimation in building systems. In: IEEE, American Control Conference, pp. 908–913.

Gunzburger, M., Peterson, J.S., Shadid, J., 2007. Reduced-order modeling of time-dependent PDEs with multiple parameters in the boundary data. Comput. Methods Appl. Mech. Eng. 196 (4–6), 1030–1047.

Haddad, W., Chellaboina, V.S., 2008. Nonlinear Dynamical Systems and Control: A Lyapunov-Based Approach. Princeton University Press, Princeton, NJ.

Huges, R., 2003. The flow of human crowds. Annu. Rev. Fluid Mech. 35, 169–182.

Juditsky, A., Hjalmarsson, H., Benveniste, A., Deylon, B., Ljung, L., Sjoberg, J., Zhang, Q., 1995. Nonlinear black-box modeling in system identification: mathematical foundations. Automatica 31 (12), 1725–1750.

Karamanos, G.S., Karniadakis, G.E., 2000. A spectral vanishing viscosity method for large-eddy simulations. J. Comput. Phys. 163 (1), 22–50.

Khalil, H., 2002. Nonlinear Systems, third ed. Prentice Hall, Englewood Cliffs, NJ.

Kitagawa, G., 1998. A self-organizing state-space model. J. Am. Stat. Assoc. 93 (443), 1203–1215.

Klein, V., Morelli, E.A., 2006. Aircraft system identification: theory and practice, AIAA Education Series. American Institute of Aeronautics and Astronautics, New York.

Kramer, B., 2011. Model reduction of the coupled burgers equation in conservation form. Masters of Science in Mathematics, Virginia Polytechnic Institute and State University.

Krstic, M., 2000. Performance improvement and limitations in extremum seeking. Syst. Control Lett. 39, 313–326.

Kunisch, K., Volkwein, S., 2007. Galerkin proper orthogonal decomposition methods for a general equation in fluid dynamics. SIAM J. Numer. Anal. 40 (2), 492–515.

Leontaritis, I., Billings, S., 1985. Input-output parametric models for non-linear systems. Part II: stochastic non-linear systems. Int. J. Control 41 (2), 329–344.

Lesieur, M., Metais, O., 1996. New trends in large-eddy simulations of turbulence. Annu. Rev. Fluid Mech. 28 (1), 45–82.

Li, K., Su, H., Chu, J., Xu, C., 2013. A fast-POD model for simulation and control of indoor thermal environment of buildings. Build. Environ. 60, 150–157.

Ljung, L., 2010. Perspectives on system identification. Autom. Remote Control 34 (1), 1–12.

Ljung, L., Glad, T., 1994. On global identifiability of arbitrary model parameterizations. Automatica 30 (2), 265–276.

Ljung, L., Vicino, A., 2005. Special issue on identification. IEEE Trans. Autom. Control 50 (10).

Ma, X., Karniadakis, G.E., 2002. A low-dimensional model for simulating three-dimensional cylinder flow. J. Fluid Mech. 458, 181–190.

MacKunis, W., Drakunov, S., Reyhanoglu, M., Ukeiley, L., 2011. Nonlinear estimation of fluid velocity fields. In: IEEE, Conference on Decision and Control, pp. 6931–6935.

Malisoff, M., Mazenc, F., 2005. Further remarks on strict input-to-state stable Lyapunov functions for time-varying systems. Automatica 41 (11), 1973–1978.

Moase, W., Manzie, C., Brear, M., 2009. Newton-like extremum seeking part I: theory. In: IEEE, Conference on Decision and Control, pp. 3839–3844.

Montseny, G., Audounet, J., Matignon, D., 1997. Fractional integro-differential boundary control of the Euler-Bernoulli beam. In: IEEE, Conference on Decision and Control, San Diego, California, pp. 4973–4978.

Narendra, K., Parthasarathy, K., 1990. Identification and control of dynamical systems using neural networks. IEEE Trans. Neural Netw. 1, 4–27.

Noack, B.R., Afanasiev, K., Morzynski, M., Tadmor, G., Thiele, F., 2003. A hierarchy of low-dimensional models for the transient and post-transient cylinder wake. J. Fluid Mech. 497, 335–363.

Noack, B., Papas, P., Monkewitz, P., 2005. The need for a pressure-term representation in empirical Galerkin models of incompressible shear flows. J. Fluid Mech. 523, 339–365.

Noack, B., Schlegel, M., Ahlborn, B., Mutschke, G., Morzynski, M., Comte, P., Tadmor, G., 2008. A finite-time thermodynamics of unsteady fluid flows. J. Non-Equilib. Thermodyn. 33 (2), 103–148.

Noack, B.R., Morzynski, M., Tadmor, G., 2011. Reduced-Order Modelling for Flow Control, first ed., vol. 528. Springer-Verlag, Wien.

Noase, W., Tan, Y., Nesic, D., Manzie, C., 2011. Non-local stability of a multi-variable extremum-seeking scheme. In: IEEE, Australian Control Conference, pp. 38–43.

Peterson, K., Stefanopoulou, A., 2004. Extremum seeking control for soft landing of electromechanical valve actuator. Automatica 40, 1063–1069.

Pillonetto, G., Dinuzzo, F., Chenc, T., Nicolao, G.D., Ljung, L., 2014. Kernel methods in system identification, machine learning and function estimation: a survey. Automatica 50, 1–12.

Rangan, S., Wolodkin, G., Poolla, K., 1995. New results for Hammerstein system identification. In: IEEE, Conference on Decision and Control. IEEE, New Orleans, USA, pp. 697–702.

Rempfer, D., 1991. Koharente strukturen und chaos beim laminar-turbulenten grenzschich-tumschlag. Ph.D. thesis, University of Stuttgart.

Rotea, M.A., 2000a. Analysis of multivariable extremum seeking algorithms. In: Proceedings of the American Control Conference, vol. 1, pp. 433–437.

Rowley, C., 2005. Model reduction for fluids using balanced proper orthogonal decomposition. Int. J. Bifurcation Chaos 15 (3), 997–1013.

San, O., Borggaard, J., 2014. Basis selection and closure for POD models of convection dominated Boussinesq flows. In: 21st International Symposium on Mathematical Theory of Networks and Systems, Groningen, The Netherlands, pp. 132–139.

San, O., Iliescu, T., 2013. Proper orthogonal decomposition closure models for fluid flows: Burgers equation. Int. J. Numer. Anal. Model. 1 (1), 1–18.

Schittkowski, K., 2002. Numerical Data Fitting in Dynamical Systems. Kluwer Academic Publishers, Dordrecht.

Schon, T., Gustafsson, F., 2003. Particle filters for system identification of state-space models linear in either parameters or states. In: Proceedings of the 13th IFAC Symposium on System Identification. IFAC, Rotterdam, The Netherlands, pp. 1287–1292.

Schön, T.B., Wills, A., Ninness, B., 2006. Maximum likelihood nonlinear system estimation. In: Proceedings of the 14th IFAC Symposium on System Identification, IFAC, Newcastle, Australia, pp. 1003–1008.

Sirisup, S., Karniadakis, G.E., 2004. A spectral viscosity method for correcting the long-term behavior of POD models. J. Comput. Phys. 194 (1), 92–116.

Sjoberg, J., Zhang, Q., Ljung, L., Benveniste, A., Deylon, B., Glorennec, P.Y., Hjalmarsson, H., Juditsky, A., 1995. Nonlinear black-box modeling in system identification: a unified overview. Automatica 31 (12), 1691–1724.

Sontag, E.D., 2008. Input to state stability: basic concepts and results. In: Nonlinear and Optimal Control Theory, Lecture Notes in Mathematics, vol. 1932. Springer-Verlag, Berlin, pp. 163–220.

Sordalen, O., 1997. Optimal thrust allocation for marine vessels. Control Eng. Pract. 15 (4), 1223–1231.

Spong, M.W., 1992. On the robust control of robot manipulators. IEEE Trans. Autom. Control 37 (11), 1782–2786.

Stankovica, M.S., Stipanovicb, D.M., 2010. Extremum seeking under stochastic noise and applications to mobile sensors. Automatica 46 (8), 1243–1251.

Tadmor, E., 1989. Convergence of spectral methods for nonlinear conservation laws. SIAM J. Numer. Anal. 26 (1), 30–44.

Tan, Y., Nesic, D., Mareels, I., 2008. On the dither choice in extremum seeking control. Automatica 44, 1446–1450.

Wang, Z., 2012. Reduced-order modeling of complex engineering and geophysical flows: analysis and computations. Ph.D. thesis, Virginia Tech.

Wang, Y., Stefanopoulou, A., Haghgooie, M., Kolmanovsky, I., Hammoud, M., 2000. Modelling of an electromechanical valve actuator for a camless engine. In: 5th International Symposium on Advanced Vehicle Control, Number 93.

Wang, Z., Akhtar, I., Borggaard, J., Iliescu, T., 2012. Proper orthogonal decomposition closure models for turbulent flows: a numerical comparison. Comput. Methods Appl. Mech. Eng. 237–240, 10–26.

CHAPTER 6

Extremum Seeking-Based Iterative Learning Model Predictive Control (ESILC-MPC)

6.1 INTRODUCTION

Model predictive control (MPC) (e.g., Mayne et al., 2000) is a model-based framework for optimal control of constrained multivariable systems. MPC is based on the repeated, receding horizon solution of a finite-time optimal control problem formulated based on the system dynamics, constraints on system states, constraints on the inputs, possibly constraints on the outputs, and a cost function describing the control objective. MPC has been applied to several applications, such as aerospace (e.g., Hartley et al., 2012) and mechatronic systems (e.g., Grancharova and Johansen, 2009). Because MPC is a model-based controller, its performance inevitably depends on the quality of the prediction model used in the optimal control computation.

In contrast, as we have seen in the previous chapters of this book, extremum–seeking (ES) control is a well-known approach where the extremum of a cost function associated with a given process performance (under some conditions) is found without the need of a detailed model of the system or the cost function (please refer to Chapter 2). Several ES algorithms and their associated stability analysis have been proposed (e.g., Krstić, 2000; Ariyur and Krstic, 2002, 2003; Tan et al., 2006; Nesic, 2009; Tan et al., 2006; Rotea, 2000; Guay et al., 2013), and many applications of ES have been reported (e.g., Zhang et al., 2003; Hudon et al., 2008; Zhang and Ordóñez, 2012; Benosman and Atinc, 2013a,b).

The idea that we want to introduce in this chapter is that the performance of a model-based MPC controller can be combined with the robustness of a model-free ES learning algorithm. This combination will lead to simultaneous identification and control of linear time–invariant systems with structural uncertainties. By identifying (or reidentifying) the system dynamics online and updating the MPC prediction model, the

Learning-Based Adaptive Control
http://dx.doi.org/10.1016/B978-0-12-803136-0.00006-3
223

closed-loop performance may be enhanced relative to a standard MPC scheme that uses an inaccurate (or outdated) model.

Many techniques and heuristics have been developed in recent years for simultaneous identification and optimization. For example, in Lobo and Boyd (1999) an approximation of dynamic programming is developed for a linear input-output map with no explicit dynamics. Approaches for more complex systems avoid dynamic programming altogether and, instead, suboptimally trade off between inputs that excite the system and inputs that regulate the state. Excitation signals are often designed to satisfy *persistency of excitation* (PE) conditions. For example, a dithering signal may be added on top of the nominal control (Sotomayor et al., 2009), although difficulties arise in determining the amplitude of the signal, and the dither indiscriminately adds noise to the process. More sophisticated schemes employ *optimal input design*, usually in the frequency domain, where maximizing the Fisher information matrix can be cast as a semidefinite program (Jansson and Hjalmarsson, 2005). However, design in the frequency domain leads to difficulties with constraints that are more naturally addressed in the time domain, for example, input (and possibly output) amplitude constraints. While the problem formulation in the time domain is highly nonconvex, developing such techniques is desirable, and thus the focus of recent work (Marafioti et al., 2014; Žáčeková et al., 2013; Rathouský and Havlena, 2013; Genceli and Nikolaou, 1996; Heirung et al., 2013; Adetola et al., 2009; González et al., 2013; Aswani et al., 2013).

In this chapter, we aim at proposing an alternative approach to realize an iterative learning-based adaptive MPC. We introduce an ES-based iterative learning MPC that merges a model-based linear MPC algorithm with a model-free ES algorithm to realize an iterative learning MPC that adapts to structured model uncertainties. This approach is an application to the MPC framework of the modular input-to-state stability (ISS)-based modular adaptive control methodology presented in this book. Due to the iterative nature of the learning model improvement, we want here to compare the proposed approach to some existing iterative learning control (ILC)-based MPC methods. Indeed, the ILC method introduced in Arimoto (1990) is a control technique which focuses on improving tracking performance of processes that repeatedly execute the same operation over time. This is of particular importance in robotics and in chemical process control of batch processes. We refer the reader to Wang et al. (2009) and Moore (1999), and Ahn et al. (2007) for more details on ILC and its applications. Tracking of reference signals subject to constraints for nonlinear systems is also an important topic that has received a lot of attention in the literature. Some of the main approaches to solve this problem are through MPC and the use of reference governors. We refer the reader to, for example, Mayne et al. (2000) and Bemporad (1998) for a detailed explanation of the two approaches. At the intersection of learning-based control and constrained control is the iterative learning-based MPC concept. ILC is a feedforward control strategy that improves the tracking error from the previous iteration and targets asymptotic tracking as the number of iterations increases. MPC is a proven control design technique for tracking/regulation with constraint satisfaction, but because there is no learning from the previous trial, MPC cannot reduce the tracking error from the knowledge of the previous trial. This motivates the ILC-based MPC schemes studied in the literature.

ILC-based MPC for chemical batch processes are studied in Wang et al. (2008), Cueli and Bordons (2008), and Shi et al. (2007). As noted in Cueli and Bordons (2008), one of the shortcomings of the current literature is a rigorous justification of feasibility, and Lyapunov-based stability analysis for ILC-based MPC. For instance, in Wang et al. (2008) the goal is to reduce the error between the reference and the output over multiple trials while satisfying only input constraints. However, the reference signals are arbitrary and the MPC scheme for tracking such signals is not rigorously justified. Furthermore, the MPC problem does not have any stabilizing conditions (terminal cost or terminal constraint set). The ILC update law is an addition of the MPC signal of the current trial to the MPC signal of the previous trial. In Cueli and Bordons (2008), an ILC-based MPC scheme for a general class of nonlinear systems with disturbances is proposed. An initial control sequence is first applied, and a linearized model is calculated. The MPC scheme is then applied and after the trial the signals of interest are filtered and a new linearized model is obtained along the state and input trajectory of the previous trial, the process is then repeated. The MPC cost function penalizes the input signal strength and the tracking error. The proof is presented only for MPC without constraints. In Shi et al. (2007), the ILC update law is designed using MPC. The cost function for the MPC problem can be defined over the same cycle or over multiple cycles. State constraints are not considered in Shi et al. (2007). In Lee et al. (1999) a batch MPC is proposed, which integrates conventional MPC scheme with an iterative learning scheme. A simplified static input-output map is considered in the paper as opposed to a dynamical system. A model for the error propagation is obtained as a function of the change in input from previous iterations, and the MPC optimization problem penalizes the tracking error and the change in input over iterations.

In summary, we think that there is a need for more rigorous theoretical justification attempted in this chapter. Furthermore, to the best of our knowledge (at the time this chapter was written), the literature on ILC-based MPC schemes does not consider state constraints nor treat robust feasibility issues in the MPC tracking problem. Rigorous justification of reference tracking proofs for the MPC is not present in the literature and stability proofs for the combination of the ILC and MPC schemes are not established in a systematic manner. Finally, we want to cite the work of Aswani et al. (2012a,b, 2013), where similar control objectives as the one targeted in this chapter have been studied using a learning-based MPC approach. The main differences are in the control/learning design methodology and the proof techniques.

In this work we use existing results on robust tracking MPC and ES learning algorithms to design a modular design approach for ILC-based MPC which ensures feasibility of the MPC during the estimation process, and also leads to improved tracking performance as the number of trials increase. We resort here to ES-based learning algorithms to improve the tracking performance as opposed to using conventional ILC update algorithms for reducing error. The motivation for this shift in approach comes from two aspects. First, in the current scheme, not only is the performance of the tracking improved over iterations through error reduction, but we also accomplish identification of the uncertain parameters of the model. This is done by optimizing an appropriate learning cost function. Second, the ES-based algorithms have been established in the literature to have certain robust convergence properties, which will be exploited in this chapter.

Indeed, in the literature two types of approaches for ES are prevalent. One is through the use of sinusoidal dither signals (see, e.g., Krstić, 2000; Krstić and Wang, 2000; Ariyur and Krstic, 2003) and the other is through the use of nonlinear programming methods (see, e.g., Khong et al., 2013a,b; Popovic et al., 2006). We first use the nonlinear programming-based algorithms for ES due to their well-established convergence results, available for the case where the uncertain parameters are bounded and the algorithm searches for parameter values that optimize certain cost functions only among such bounded domains. This property will be crucial to establishing robust feasibility of the MPC algorithm during the parameter estimation process. In a second step, we simplify the results by using a dither-based multiparametric ES (MES) learning approach, together with a simplified MPC (without guarantees of robust feasibility during the learning process). The later algorithm is more targeted toward real-time applications, where the computation power is usually limited.

In this chapter we also restrict the reference signals for MPC tracking to piecewise constant trajectories. See Limon et al. (2008) and Ferramosca et al. (2009) for asymptotic tracking of piecewise constant reference signal for linear constrained systems. These results are further extended to MPC tracking for linear constrained systems with bounded additive disturbances in Limon et al. (2010) and Alvarado et al. (2007a). MPC for tracking periodic reference signals is presented in Limon et al. (2012).

In brief, the main contribution of this chapter is to present a rigorous proof of an ES-based iterative learning control (ESILC)-based MPC scheme using existing Lyapunov function-based stability analysis established in Limon et al. (2010), and ES algorithms in Khong et al. (2013b), to justify the modular design method for ILC-based MPC proposed in Benosman et al. (2014), where a modular approach to design ILC-based MPC schemes for a class of constrained linear systems was proposed.

We consider the case where the true model of the plant contains parametric uncertainties in the system's matrices, and the goal is to track a reference trajectory while satisfying certain state and input constraints. The approach is first to design an MPC scheme with an estimated model of the plant, such that an ISS property with respect to the parameter estimation error is guaranteed, and further choose a robust ES algorithm to guarantee convergence of the model uncertainties' estimates to the true parameters' values. After every trial the model of the estimated plant in the MPC scheme is updated and tracking is performed. Intuitively, as the number of trials increases the error between the true parameter and estimated parameters decreases due to the ES algorithm, and the ISS property implies that tracking performance improves over time. In this chapter we provide a theoretical justification for this ESILC-based MPC scheme.

The rest of the chapter is organized as follows. Section 6.2 contains some of the notations and definitions used throughout the rest of the chapter (for more details about the mathematical tools please refer to Chapter 1). The MPC problem formulation is presented in Section 6.3. Section 6.4 is dedicated to a rigorous analysis of a nonlinear programming ES-based iterative MPC. In Section 6.5, we present a simplified solution which uses a dither-based ES control to write a simplified iterative adaptive

MPC algorithm. Finally, simulation results and concluding comments are presented in Sections 6.6 and 6.7, respectively.

6.2 NOTATION AND BASIC DEFINITIONS

Throughout this chapter, $\mathbb{R}$ denotes the set of real numbers and $\mathbb{Z}$ denotes the set of integers. State constraints and input constraints are represented by $\mathcal{X} \subset \mathbb{R}^n$ and $\mathcal{U} \subset \mathbb{R}^m$, respectively. The optimization horizon for MPC is denoted by $N \in \mathbb{Z}_{\geq 1}$. The feasible region for the MPC optimization problem is denoted by $\mathcal{X}_N$. A continuous function $\alpha: \mathbb{R}_{\geq 0} \to \mathbb{R}_{\geq 0}$ with $\alpha(0) = 0$ belongs to class $\mathcal{K}$ if it is increasing and bounded. A function β belongs to class $\mathcal{K}_\infty$ if it belongs to class $\mathcal{K}$ and is unbounded. A function $\beta(s, t) \in \mathcal{KL}$ if $\beta(\cdot, t) \in \mathcal{K}$ and $\lim_{t \to \infty} \beta(s, t) = 0$. Given two sets A and B, such that $A \subset \mathbb{R}^n$, $B \subset \mathbb{R}^n$, the Minkowski sum is defined as $A \oplus B := \{a + b | a \in A, b \in B\}$. The Pontryagin set difference is defined as $A \ominus B := \{x | x \oplus B \in A\}$. Given a matrix $M \in \mathbb{R}^{m \times n}$, the set $MA \subset \mathbb{R}^m$ is defined as $MA \triangleq \{Ma: a \in A\}$. A positive definite matrix is denoted by $P > 0$. The standard Euclidean norm is represented as $\|x\|$ for $x \in \mathbb{R}^n$, $\|x\|_P := \sqrt{x^T P x}$ for a positive definite matrix P, $\|x\|_\mathcal{A} := \inf_{y \in \mathcal{A}} \|x - y\|$ for a closed set $\mathcal{A} \subset \mathbb{R}^n$ and $\|A\|$ represents an appropriate matrix norm where A is a matrix. $\mathbb{B}$ represents the closed unit ball in the Euclidean space. Also, a matrix $M \in \mathbb{R}^{n \times n}$ is said to be Schur iff all its eigenvalues are inside the unitary disk. Finally, we denote the ith element of a vector y, by y_{-i}.

In the next section we will describe in detail the problem studied in this chapter.

6.3 PROBLEM FORMULATION

We consider linear systems of the form

$$x(k + 1) = (A + \Delta A)x(k) + (B + \Delta B)u(k), \tag{6.1}$$

$$y(k) = Cx(k) + Du(k), \tag{6.2}$$

where ΔA and ΔB represent the uncertainty in the system model. We will assume that the uncertainties are bounded.

Assumption 6.1. *The uncertainties* $\|\Delta A\| \leq \ell_A$ *and* $\|\Delta B\| \leq \ell_B$ *for some* $\ell_A, \ell_B > 0$.

Next, we impose some assumptions on the reference signal r.

Assumption 6.2. *The reference signal* $r: [0, T] \to \mathbb{R}$ *is a piecewise constant trajectory for some* $T > 0$.

Under Assumptions 6.1 and 6.2, the goal is to design a control scheme guaranteeing tracking with sufficiently small errors by learning the uncertain parameters of the system. We will explain in detail the optimization problem associated with the MPC-based controller. The results stated here are from Limon et al. (2010). We exploit the analysis results in Limon et al. (2010) to establish that the closed-loop system has an ISS property with respect to the parameter estimation error.

Because the value of ΔA and ΔB are not known a priori, the MPC uses a model of the plant based on the current estimate $\hat{\Delta} A$ and $\hat{\Delta} B$.

We will now formulate the MPC problem with a given estimate of the uncertainty for a particular iteration of the learning process. We rewrite the system dynamics as

$$x(k+1) = f(x, u) + g(x, u, \Delta) = F(x, u, \Delta), \tag{6.3}$$

where $f(x, u) = Ax + Bu$ and $g(x, u, \Delta) = \Delta Ax + \Delta Bu$.

Assumption 6.3. *The state constraint set $\mathcal{X} \subset \mathbb{R}^n$ and control constraint set $\mathcal{U} \subset \mathbb{R}^m$ are compact, convex polyhedral sets.*

The MPC model is generated using an estimate $\hat{\Delta} A$, $\hat{\Delta} B$ and is expressed as

$$x(k+1) = f(x, u) + g(x, u, \hat{\Delta}) = F(x, u, \hat{\Delta}). \tag{6.4}$$

We can now rewrite the actual model as

$$x(k+1) = f(x, u) + g(x, u, \hat{\Delta}) + (\Delta A - \hat{\Delta} A)x + (\Delta B - \hat{\Delta} B)u. \tag{6.5}$$

This system can now be compared to the model in Limon et al. (2010), in which case we write it as

$$x(k+1) = F(x(k), u(k), \hat{\Delta}) + w(k), \tag{6.6}$$

where

$$w(k) = (\Delta A - \hat{\Delta} A)x(k) + (\Delta B - \hat{\Delta} B)u(k), \tag{6.7}$$

and $x(k) \in \mathcal{X}, u(k) \in \mathcal{U}$. The following assumption will be explained in more detail in the next section.

Assumption 6.4. *The estimates of the uncertain parameters are bounded with $\|\hat{\Delta} A\| \le \ell_A$, and $\|\hat{\Delta} B\| \le \ell_B$ for all iterations of the ES algorithm.*

Remark 6.1. *Assumption 6.4 is reasonable if the choice of the ES algorithm restricts the search of the true parameter values to the prescribed domain of the uncertainty. For example, the DIRECT algorithm (e.g., Jones et al., 1993) restricts the search for the minimizer to compact (multidimensional) intervals. We also refer* *the reader to Khong et al. (2013b, Section 3) for other algorithms satisfying such boundedness property.*

We now impose certain conditions on the disturbance $w(k)$ and the system's matrices in accordance with Limon et al. (2010, Assumption 1).

Assumption 6.5. *The pair $(A + \hat{\Delta}A, B + \hat{\Delta}B)$ is controllable for every realization of $\hat{\Delta}A$ and $\hat{\Delta}B$.*

Remark 6.2. *It follows from Eqs. (6.3), (6.1), (6.4) that the disturbance w is bounded. So the disturbance belongs to the set $\mathcal{W} := \{w: |w| \le w^*\}$ for some w^* (which depends on $\mathcal{X}, \mathcal{U}, \Delta A - \hat{\Delta}A, \Delta B - \hat{\Delta}B$). Because we assume full state measurement, all the conditions of Limon et al. (2010, Assumption 1) are satisfied.*

We will denote the actual model using (x, u) and the MPC model through $(\bar{x}, \bar{u})$. Hence we have

$$x(k+1) = F(x, u, \hat{\Delta}) + w,$$
$$\bar{x}(k+1) = F(\bar{x}, \bar{u}, \hat{\Delta}).$$

6.3.1 Robust Positive Invariant Sets

We denote the error between the states of the true model and the MPC model by $e(k) = x(k) - \bar{x}(k)$. We want the error to be bounded during tracking. The error dynamics is then given by

$$e(k+1) = (A + \hat{\Delta}A + (B + \hat{\Delta}B)K)e(k) + w(k), \qquad (6.8)$$

where $u = \bar{u} + Ke$ and the matrix K is such that $A_K := (A + \hat{\Delta}A + (B + \hat{\Delta}B)K)$ is Schur.

We first recall the definition of a robust positive invariant (RPI) set, see for example, Limon et al. (2010) (refer also to Chapter 1).

Definition 6.1. *A set Φ_K is called an RPI set for the uncertain dynamics (6.8), if $A_K\Phi_k \oplus \mathcal{W} \subseteq \Phi_K$.*

We let Φ_K be an RPI set associated with the error dynamics (6.8), that is, $A_K\Phi_K \oplus \mathcal{W} \subseteq \Phi_K$.

Remark 6.3. *An example of algorithms calculating RPI sets can be found in Borrelli et al. (2012). From our assumptions this algorithm terminates in a finite number of steps because A_K is Schur, $\mathcal{X}, \mathcal{U}, \mathcal{W}$ are bounded, and the error dynamics is linear.*

6.3.2 Tightening the Constrains

Now we follow Limon et al. (2010) and tighten the constraints for the MPC model so that we achieve robust constraint satisfaction for the actual model with uncertainties. Let $\mathcal{X}_1 = X \ominus \Phi_K$, and $\mathcal{U}_1 = \mathcal{U} \ominus K\Phi_K$. The following result is from Alvarado et al. (2007b, Proposition 1, Theorem 1, and Corollary 1).

Proposition 6.1. *Let Φ_K be RPI for the error dynamics. If $e(0) \in \Phi_K$, then $x(k) \in \bar{x}(k) \oplus \Phi_K$ for all $k \geq 0$ and $w(k) \in \mathcal{W}$. If in addition, $\bar{x}(k) \in \mathcal{X}_1$ and $\bar{u}(k) \in \mathcal{U}_1$ then with the control law $u = \bar{u} + Ke$, $x(k) \in \mathcal{X}$, and $u(k) \in \mathcal{U}$ for all $k \geq 0$.*

6.3.3 Invariant Set for Tracking

As in Alvarado et al. (2007b) and Limon et al. (2010), we will characterize the set of nominal steady states, and inputs so that we can relate them later to the tracking problem. Let $z_s = (\bar{x}_s, \bar{u}_s)$ be the steady state for the MPC model. Then,

$$\begin{bmatrix} A + \hat{\Delta}A - I & B + \hat{\Delta}B \\ C & D \end{bmatrix} \begin{bmatrix} \bar{x}_s \\ \bar{u}_s \end{bmatrix} = \begin{bmatrix} 0 \\ \bar{y}_s \end{bmatrix}. \tag{6.9}$$

From the controllability assumption on the system matrices, the admissible steady states can be characterized by a single parameter $\bar{\theta}$ as

$$\bar{z}_s = M_\theta \bar{\theta}, \tag{6.10}$$

$$\bar{y}_s = N_\theta \bar{\theta}, \tag{6.11}$$

for some $\bar{\theta}$, matrices M_θ and $N_\theta = [C\ D]M_\theta$. We let $\mathcal{X}_s, \mathcal{U}_s$ denote the set of admissible steady states that are contained in $\mathcal{X}_1, \mathcal{U}_1$ and satisfy Eq. (6.9). $\mathcal{Y}_s$ denotes the set of admissible output steady states. Now we will define an invariant set for tracking which will be utilized as a terminal constraint for the optimization problem.

Definition 6.2 (Limon et al., 2010, Definition 2). *An invariant set for tracking for the MPC model is the set of initial conditions, steady states and inputs (characterized by $\bar{\theta}$) that can be stabilized by the control law $\bar{u} = \bar{K}\bar{x} + L\bar{\theta}$ with $L := [-\bar{K}\ I]M_\theta$ while $(\bar{x}(k), \bar{u}(k)) \in \mathcal{X}_1 \times \mathcal{U}_1$ for all $k \geq 0$.*

We choose the matrix $\bar{K}$ such that $A_{\bar{K}} := (A + \hat{\Delta}A + (B + \hat{\Delta}B)\bar{K})$ is Schur. We refer the reader to Alvarado et al. (2007b) and Limon et al. (2010) for more details on computing the invariant set for tracking. We will refer to the invariant set for tracking as $\Omega_{\bar{K}}$. We say a point $(\bar{x}(0), \bar{\theta}) \in \Omega_{\bar{K}}$ if with the control law $u = \bar{K}(\bar{x} - \bar{x}_s) + \bar{u}_s = \bar{K}\bar{x} + L\bar{\theta}$, the solutions of the MPC model from $\bar{x}(0)$ satisfy $\bar{x}(k) \in \text{Proj}_x(\Omega_{\bar{K}})$ for all $k \geq 0$. As stated in Limon et al. (2010) the set can be taken to be a polyhedral.

6.3.4 MPC Problem

Now we will define the optimization problem that will be solved at every instant to determine the control law for the actual plant dynamics. For a given target set point y_t and initial condition x, the optimization problem $\mathcal{P}_N(x, y_t)$ is defined as,

$$\min_{\bar{x}(0), \bar{\theta}, \bar{\mathbf{u}}} \quad V_N(x, y_t, \bar{x}(0), \bar{\theta}, \bar{\mathbf{u}}),$$

$$\text{s.t.} \quad \bar{x}(0) \in x \oplus (-\Phi_K),$$

$$\bar{x}(k+1) = (A + \hat{\Delta}A)\bar{x}(k) + (B + \hat{\Delta}B)\bar{u}(k),$$

$$\bar{x}_s = M_\theta \bar{\theta},$$

$$\bar{y}_s = N_\theta \bar{\theta},$$

$$(\bar{x}(k), \bar{u}(k)) \in \mathcal{X}_1 \times \mathcal{U}_1, k \in \mathbb{Z}_{\leq N-1},$$

$$(\bar{x}(N), \bar{\theta}) \in \Omega_{\bar{K}},$$

where the cost function is defined as follows:

$$V_N(x, y_t, \bar{x}(0), \bar{\theta}, \bar{\mathbf{u}}) = \sum_{k=0}^{N-1} \|\bar{x}(k) - \bar{x}_s\|_Q^2$$

$$+ \|\bar{u}(k) - \bar{u}_s\|_R^2 + \|\bar{x}(N) - \bar{x}_s\|_P^2 + \|\bar{y}_s - y_t\|_T^2.$$

$$(6.12)$$

Such cost function is frequently used in MPC literature for tracking, except for the additional term in the end which penalizes the difference between the artificial steady state and the actual target value. We refer the reader to Alvarado et al. (2007a,b), and Limon et al. (2010) for more details.

Remark 6.4. *As it can be observed, the MPC model for optimization, the terminal set for tracking, and the robust positively invariant set are all dependent on the current estimate of the uncertainty $\hat{\Delta}A$ and $\hat{\Delta}B$. Hence, as the ES algorithm updates the estimates, the optimization problem must be recalculated for the new values of $\hat{\Delta}A$ and $\hat{\Delta}B$. This is a potential disadvantage of this version of the ESILC-MPC scheme.*

Assumption 6.6. *The following conditions are satisfied by the optimization problem.*
1. *The matrices $Q > 0, R > 0, T > 0$.*
2. *$(A + \hat{\Delta}A + (B + \hat{\Delta}B)K)$ is Schur matrix, Φ_K is an RPI set for the error dynamics, and $\mathcal{X}_1, \mathcal{U}_1$ are nonempty.*

3. *The matrix* $\bar{K}$ *is such that* $A + \hat{\Delta}A + (B + \hat{\Delta}B)\bar{K}$ *is Schur and* $P > 0$ *satisfies*

$$P - (A + \hat{\Delta}A + (B + \hat{\Delta}B)\bar{K})^T P(A + \hat{\Delta}A + (B + \hat{\Delta}B)\bar{K}) = Q + \bar{K}^T R\bar{K}.$$

4. *The set* $\Omega_{\bar{K}}$ *is an invariant set for tracking subject to the tightened constraints* $\mathcal{X}_1, \mathcal{U}_1$.

As noted in Limon et al. (2010), the feasible set $\mathcal{X}_N$ does not vary with the set points γ_t and the optimization problem is a quadratic programming problem. The optimal values are given by $\bar{x}_s^*, \bar{u}^*(0), \bar{x}^*$. The MPC law writes then as $u = \kappa_N(x) = K(x - \bar{x}^*) + \bar{u}^*(0)$. The MPC law κ_N implicitly depends on the current estimate of the uncertainty $\hat{\Delta}$. Also it follows from the results in Bemporad et al. (2002) that the control law for the MPC problem is continuous.

Remark 6.5. *The reference trajectory consists of a collection of set points* $\{\gamma_j\}_{j=0}^{j^*}$. *The optimization is set up with respect to a set point as opposed to the reference trajectory because it is noted in Limon et al. (2010, Property 1) that the optimization problem remains feasible even when the set points are changing for the same constraints/terminal set/controller gains. In Limon et al. (2010) because the tracking is only asymptotic, an implicit assumption is that the set points can only be changed after sufficiently long times to maintain good reference tracking.*

6.4 THE DIRECT ES-BASED ITERATIVE LEARNING MPC

6.4.1 DIRECT-Based Iterative Learning MPC

In this section we will explain the assumptions regarding the learning cost function[1] used for identifying the true parameters of the uncertain system via nonlinear programming-based ES. Let Δ be a vector that contains the entries in ΔA and ΔB. Similarly, the estimate will be denoted by $\hat{\Delta}$. Then $\Delta, \hat{\Delta} \in \mathbb{R}^{n(n+m)}$.

Because we do not impose the presence of attractors for the closed-loop system as in Popovic et al. (2006) or Khong et al. (2013a), the cost function that we use here, $Q: \mathbb{R}^{n(n+m)} \rightarrow \mathbb{R}_{\geq 0}$, depends on x_0. For iterative learning methods, the same initial condition x_0 is used to learn the uncertain parameters, and hence we refer to $Q(x_0, \hat{\Delta})$ as only $Q(\hat{\Delta})$ because x_0 is fixed.

---

[1] Not to be confused with the MPC cost function.

Assumption 6.7. *The learning cost function* Q: $\mathbb{R}^{n(n+m)} \rightarrow \mathbb{R}_{\geq 0}$ *is*
1. *Lipschitz in the compact set of uncertain parameters.*
2. *The true parameter* Δ *is such that* $Q(\Delta) < Q(\hat{\Delta})$ *for all* $\hat{\Delta} \neq \Delta$.

One example of a learning cost function is an identification–type cost function, where the error between outputs' measurements from the system are compared to the MPC model outputs. Another example of a learning cost function can be a performance–type cost function, where a measured output of the system is directly compared to a desired reference trajectory. Later in Section 6.6, we will test both types of learning cost functions.

We then use the DIRECT optimization algorithm introduced in Jones et al. (1993) for finding the global minimum of a Lipschitz function without explicit knowledge of the Lipschitz constant. The algorithm is implemented in MATLAB using Finkel (2003). We will utilize a modified termination criterion introduced in Khong et al. (2013a) for the DIRECT algorithm to make it more suitable for ES applications. As we will mention in later sections, the DIRECT algorithm has nice convergence properties which will be used to establish our main stability results.

In Jones et al. (1993), it is stated that the DIRECT algorithm works as long as Q is continuous in a neighborhood around the minimum Δ, although in other works, such as Khong et al. (2013a,b), it is claimed that DIRECT algorithms require Lipschitz continuity. We make here a stronger assumption on Q to avoid any inconsistency with the literature on ES.

We want to point out here that DIRECT is a sampling–based optimization algorithm. In order to arrive at an updated $\Delta(k + 1)$, the uncertainty domain will be probed many times. In essence, many function evaluations (probes) of the cost function Q are necessary for one iterative update of the estimate Δ. In order to simplify the presentation, we have not included this aspect in the algorithm. This presentation is also consistent with the implementation scheme in Khong et al. (2013b). The initial trial point Δ_0 is usually given by the midpoint of the interval bounds for the uncertain parameters.

Remark 6.6. *To compare the ILC-based MPC scheme proposed here to some of the features of traditional ILC stated in Moore (1999), we can note the following: the ILC-based MPC scheme is similar to other ILC approaches as the initial conditions are reset after every trial, the trial duration is finite and is characterized by $T > 0$, the learning after a completed trial is model-free (we will later show through detailed proofs that the tracking error becomes smaller as the number of iterations grows), and the learning process is independent of the desired reference trajectory. On the other hand, we also learn the true process parameters as opposed to just reducing the error between trials. Also we use an MPC-based* *scheme for regulation/tracking purposes, as most ILC schemes are directly applicable. The update law for the ILC-based MPC scheme is not explicitly mentioned as in other ILC schemes because the update of the estimated parameter by the DIRECT algorithm leads to an update of the control law for the next trial, as κ_N depends on the current estimate.*

Remark 6.7. *Most of the existing nonlinear program-based ES methods impose the existence of an attractor for the system dynamics. See Khong et al. (2013b) and Teel and Popovic (2001) for details. In this chapter we do not insist on the existence of such attractors, as we reset to the same initial condition after finite time over multiple iterations. Also because we deal with a tracking problem, the cost function takes into account the behavior of states only up to a finite time (associated with the length of the reference signal), as opposed to cost functions that are defined in the limit as time becomes infinite in the regulation problems in Teel and Popovic (2001) and Khong et al. (2013b).*

6.4.2 Proof of the MPC ISS-Guarantee and the Learning Convergence

We will now present the stability analysis of ESILC–MPC Algorithm 6.1, using the existing results for MPC tracking and DIRECT algorithm established in Limon et al. (2010) and Khong et al. (2013b), respectively.

First, we define the value function $V_N^*(x, y_t) = \min_{\bar{x}(0),\theta,\bar{u}} V_N(x, y_t, \bar{x}(0), \theta, \bar{u})$ for a fixed target y_t. Also, we let $\tilde{\theta} := \arg\min_{\bar{\theta}} \|N_\theta \bar{\theta} - y_t\|$,

---

ALGORITHM 6.1   DIRECT ESILC-MPC

Require: $r: [0, T] \to \mathbb{R}$, x_0, Q, Δ_0, N_{es} (number of learning iterations), Q_{th} (learning cost threshold)

1: Set $k = 0, j = 0$
2: Initial trial point of DIRECT: Δ_0
3: Form MPC model with Δ_0
4: **while** $k \leq N_{es}$ or $Q > Q_{th}$ **do**
5:     Set $x(0) = x_0, j = 0$
6:     **for** $j \leq T$ **do**
7:         Compute MPC law at $x(j)$
8:         $j = j + 1$
9:     **end for**
10:    Compute $Q(\Delta_k)$ from $\{x(j)\}_{j=0}^{T}$
11:    Find Δ_{k+1} using DIRECT and $\{Q(\Delta_j)\}_{j=0}^{k}$
12:    Update MPC model and $\mathcal{P}_N$ using Δ_{k+1}
13:    $k = k + 1$, loop to 4
14: **end while**

$(\tilde{x}_s, \tilde{u}_s) = M_\theta \tilde{\theta}$, and $\tilde{y}_s = C\tilde{x}_s + D\tilde{u}_s$. If the target steady-state y_t is not admissible, the MPC tracking scheme drives the output to converge to the point $\tilde{y}_s$ which is a steady-state output that is admissible and also minimizes the error with the target steady state, that is, graceful performance degradation principle (e.g., Benosman and Lum, 2009). The proof of the following result follows from Limon et al. (2010, Theorem 1), and classically uses $V_N^*(x, y_t)$ as a Lyapunov function for the closed-loop system.

Proposition 6.2. *Let y_t be given, then for all $x(0) \in \mathcal{X}_N$, the MPC problem is recursively feasible, the state $x(k)$ converges to $\tilde{x}_s \oplus \Phi_K$, and the output $y(k)$ converges to $\tilde{y}_s \oplus (C + DK)\Phi_K$.*

The next result states the convergence properties of the modified DIRECT algorithm, which we will use in establishing the main result. This result is stated as Khong et al. (2013b, Assumption 7), and it follows from the analysis of the modified DIRECT algorithm in Khong et al. (2013a).

Proposition 6.3. *For any sequence of updates $\hat{\Delta}$ from the modified DIRECT algorithm and $\varepsilon > 0$, there exists an $N > 0$ such that $\|\Delta - \hat{\Delta}_k\| \le \varepsilon$ for $k \ge N$.*

Remark 6.8. *Note that the results in Khong et al. (2013a) also include a robustness aspect of the DIRECT algorithm. This can be used to account for measurement noises and computational error associated with the learning cost Q.*

We now state the main result of this chapter, which combines the ISS MPC formulation and the ES algorithm.

Theorem 6.1. *Under Assumptions 6.1–6.7, given an initial condition x_0, an output target y_t, for every $\varepsilon > 0$, there exists a $T, N_1, N_2 > 0$ such that $\|y(k) - \tilde{y}_s\| \le \varepsilon$ for $k \in [N_1, T]$ after N_2 iterations (or resets) of the ESILC-MPC algorithm.*

Proof. It can observed that because the size of Φ_K grows with the size of $\mathcal{W}$ and $\Phi_K = \{0\}$ for the case without disturbances, without loss of generality $\Phi_K \subseteq \Gamma(w^*)\mathbb{B}$, where $\Gamma \in \mathcal{K}$ and $w^* = \|\Delta A - \hat{\Delta} A\|X^* + \|\Delta B - \hat{\Delta} B\|U^*$, where $X^* = \max_{x \in \mathcal{X}} \|x\|$ and $U^* = \max_{u \in \mathcal{U}} \|u\|$. Here X^*, U^* are fixed over both (regular) time and learning iteration number, but the uncertainties vary over iterations because of the modified DIRECT algorithm updates. Because the worst-case disturbance depends directly on the estimation error, without loss of generality we have that $\Phi_K \subseteq \gamma(\|\Delta - \hat{\Delta}\|)\mathbb{B}$ and $(C + DK)\Phi_K \subseteq \gamma^*(\|\Delta - \hat{\Delta}\|)\mathbb{B}$ for some $\gamma, \gamma^* \in \mathcal{K}$. It follows from Proposition 6.2 that $\lim_{k \to \infty} |x(k)|_{\tilde{x}_s \oplus \Phi_K} = 0$.

Then,

$$\lim_{k \to \infty} \|x(k) - \tilde{x}_s\| \le \max_{x \in \Phi_K} \|x\|$$

$$\le \gamma(\|\Delta - \hat{\Delta}\|).$$

We observe that the previous set of equations states that the closed-loop system with the MPC controller has the asymptotic gain property and it is upper-bounded by the size of the parameter estimation error. Note that the estimate $\hat{\Delta}$ is constant for a particular iteration of the process. Also, for the case of no uncertainties we have 0-stability (Lyapunov stability for the case of 0-uncertainty). This can be proven by using the cost function $V_N^*(x, y_t)$ as the Lyapunov function, such that $V_N^*(x(k+1), y_t) \leq V_N^*(x(k), y_t)$ and $\lambda_{\min}(Q)\|x - \tilde{x}_s\|^2 \leq V_N^*(x, y_t) \leq \lambda_{\max}(P)\|x - \tilde{x}_s\|^2$, see Limon et al. (2008). Furthermore, here the stability and asymptotic gain property can be interpreted with respect to the compact set $\mathcal{A} := \{\tilde{x}_s\}$.

Because the MPC law is continuous, the closed-loop system for a particular iteration of the ESILC-MPC scheme is also continuous with respect to the state. Then, from Cai and Teel (2009, Theorem 3.1) we can conclude that the closed-loop system is ISS with respect to the parameter estimation error and hence satisfies

$$\|x(k) - \tilde{x}_s\| \leq \beta(\|x(0) - \tilde{x}_s\|, k) + \hat{\gamma}(\|\Delta - \hat{\Delta}\|),$$

where $\beta \in \mathcal{KL}$ and $\hat{\gamma} \in \mathcal{K}$. Now, let $\varepsilon_1 > 0$ be small enough such that $\hat{\gamma}(\varepsilon_1) \leq \varepsilon/2$. From Proposition 6.3, it follows that there exists an $N_2 > 0$ such that $\|\Delta - \hat{\Delta}_t\| \leq \varepsilon_1$ for $t \geq N_2$, where t is the iteration number of the ESILC-MPC scheme. Hence there exists $N_1 > 0$ such that $\|\beta(|x(0) - \tilde{x}_s|, k)\| \leq \varepsilon/2$ for $k \geq N_1$. We choose T such that $T > N_1$. Then, we have that for $k \in [N_1, T]$ and for $t \geq N_2$,

$$\|x(k) - \tilde{x}_s\| \leq \varepsilon.$$

Similarly, using the linearity dependence between y and x, we can also establish that $\exists \, \tilde{\varepsilon}(\varepsilon)$, such that for $k \in [N_1, T]$ and for $t \geq N_2$

$$\|y(k) - \tilde{y}_s\| \leq \tilde{\varepsilon}(\varepsilon).$$

Remark 6.9. *Alternatively, Ferramosca et al. (2009, Remark 4) claim ISS property for the nominal version of the tracking problem. In this chapter we refer to the robust version of the MPC tracking problem in Limon et al. (2010) to ensure robust feasibility and arrive at an ISS result for the closed-loop system. For the case where the set points are varying as in a piecewise constant reference, the change in set points can be made after sufficiently large times using the result in Theorem 6.1. For example, given a tracking accuracy of $\varepsilon > 0$, we can use the results of Theorem 6.1 to arrive at a time $T \geq N_1$ after which the set points can be changed. The duration of the reference signal is taken to be sufficiently large. This process can be repeated in a similar way for multiple set point changes. This type of assumption on the duration* *of the tracking signal is reasonable in ILC methods for slow processes, for example, chemical batch systems.*

Remark 6.10. *Note that, although we are implicitly doing parametric identification in closed loop, we have not explicitly enforced any PE condition on the MPC. We argue that the PE is implicitly enforced, due to the search nature of the DIRECT ES algorithm, which probes the system in many directions (in the space of uncertainties) to reach the best value of the parameters. However, to be more rigorous, it is "easy" to enforce a PE condition on the MPC algorithm by adding extra constraints on the input variables, that is, PE conditions on a vector of past input values over a finite time interval (e.g., Marafioti et al., 2014).*

6.5 DITHER MES-BASED ADAPTIVE MPC

Similar to the previous section, we want to design here an adaptive controller that solves regulation and tracking problems for linear time–invariant systems with structural model uncertainties, under state constraints, input constraints, and output constraints. The difference here is that we will not be too conservative with the MPC formulation; that is, we will not guarantee robust feasibility of the MPC problem. Instead, we want to present a simple way of connecting a dither-based MES algorithm with a nominal MPC algorithm. We believe that this simplification is needed to be able to implement such ILC-MPC algorithms in real time, where the computation power is limited.

In what follows, we first present the nominal MPC problem, that is, without model uncertainties, and then extend this nominal controller to its adaptive form by merging it with a dither-based MES algorithm.

6.5.1 Constrained Linear Nominal MPC

Consider the nominal linear prediction model

$$x(k + 1) = Ax(k) + Bu(k), \qquad (6.13a)$$

$$y(k) = Cx(k) + Du(k), \qquad (6.13b)$$

where $x \in \mathbb{R}^n$, $u \in \mathbb{R}^m$, and $y \in \mathbb{R}^p$ are the state, input, and output vectors subject to constraints

$$x_{\min} \leq x(k) \leq x_{\max}, \qquad (6.14a)$$

$$u_{\min} \leq u(k) \leq u_{\max}, \qquad (6.14b)$$

$$y_{\min} \leq y(k) \leq y_{\max}, \qquad (6.14c)$$

where $x_{\min}, x_{\max} \in \mathbb{R}^n$, $u_{\min}, u_{\max} \in \mathbb{R}^m$, and $y_{\min}, y_{\max} \in \mathbb{R}^p$ are the lower and upper bounds on the state, input, and output vectors, respectively. At every control cycle $k \in \mathbb{Z}_{0+}$, MPC solves the finite horizon optimal control problem

$$\min_{U(k)} \sum_{i=0}^{N-1} \|x(i|k)\|_{Q_M}^2 + \|u(i|k)\|_{R_M}^2 + \|x(N|k)\|_{P_M}^2, \tag{6.15a}$$

$$\text{s.t.} \quad x(i+1|k) = Ax(i|k) + Bu(i|k), \tag{6.15b}$$

$$y(i|k) = Cx(i|k) + Du(i|k), \tag{6.15c}$$

$$x_{\min} \leq x(i|k) \leq x_{\max}, \quad i \in \mathbb{Z}_{[1,N_c]}, \tag{6.15d}$$

$$u_{\min} \leq u(i|k) \leq u_{\max}, \quad i \in \mathbb{Z}_{[0,N_{cu}-1]}, \tag{6.15e}$$

$$y_{\min} \leq y(i|k) \leq y_{\max}, \quad i \in \mathbb{Z}_{[0,N_c]}, \tag{6.15f}$$

$$x(0|k) = x(k), \tag{6.15g}$$

where $Q_M \geq 0$, $P_M, R_M > 0$ are symmetric weight matrices of appropriate dimensions, $N_{cu} \leq N$, $N_c \leq N - 1$ are the input and output constraint horizons along which the constraints are enforced, $N_u \leq N$ is the control horizon, and N is the prediction horizon. The MPC performance criterion is defined by Eq. (6.15a), and Eqs. (6.15d)–(6.15f) enforce the constraints. The optimization vector is $U(k) = [u'(0|k) \dots u'(N_u - 1|k)]' \in \mathbb{R}^{N_u m}$.

At time k, the MPC problem (6.15) is initialized with the current state value $x(k)$ by Eq. (6.15g) and solved to obtain the optimal sequence $U^*(k)$. Then, the input $u(k) = u_{\mathrm{MPC}}(k) = u^*(0|k) = [I_m \ 0 \ \dots \ 0]U(k)$ is applied to the system.

6.5.2 MES-Based Adaptive MPC Algorithm

Consider now the system (6.13), with structural uncertainties, such that

$$x(k+1) = (A + \Delta A)x(k) + (B + \Delta B)u(k), \tag{6.16a}$$

$$y(k) = (C + \Delta C)x(k) + (D + \Delta D)u(k), \tag{6.16b}$$

with the following assumptions.

Assumption 6.8. *The constant uncertainty matrices $\Delta A, \Delta B, \Delta C$, and ΔD, are bounded, such that $\|\Delta A\|_2 \leq l_A$, $\|\Delta B\|_2 \leq l_B$, $\|\Delta C\|_2 \leq l_C$, and $\|\Delta D\|_2 \leq l_D$, with l_A, l_B, l_C, and $l_D > 0$.*

Assumption 6.9. *There exist nonempty convex sets $\mathcal{K}_a \subset \mathbb{R}^{n \times n}$, $\mathcal{K}_b \subset \mathbb{R}^{n \times m}$, $\mathcal{K}_c \subset \mathbb{R}^{p \times n}$, and $\mathcal{K}_d \subset \mathbb{R}^{p \times m}$, such that $A + \Delta A \in \mathcal{K}_a$ for all ΔA such that $\|\Delta A\|_2 \leq l_A$, $B + \Delta B \in \mathcal{K}_b$ for all ΔB such that $\|\Delta B\|_2 \leq l_B$,*

$C + \Delta C \in \mathcal{K}_c$ for all ΔC such that $\|\Delta C\|_2 \leq l_C$, $D + \Delta D \in \mathcal{K}_d$ for all ΔD such that $\|\Delta D\|_2 \leq l_D$.

Assumption 6.10. *The iterative learning MPC problem (6.15) (and the associated reference tracking extension), where we substitute the model with structural uncertainty (6.16) for the nominal model (6.15b), (6.15c), is a well-posed optimization problem for any matrices $A + \Delta A \in \mathcal{K}_a$, $B + \Delta B \in \mathcal{K}_b$, $C + \Delta C \in \mathcal{K}_c$, and $D + \Delta D \in \mathcal{K}_d$.*

Under these assumptions, we proceed as follows: We solve the iterative learning MPC problem (6.15), where we substitute Eq. (6.16) for Eqs. (6.15b), (6.15c), iteratively such that at each new iteration we update our knowledge of the uncertain matrices ΔA, ΔB, ΔC, and ΔD, using a model-free learning algorithm, which in our case is the ES algorithm. Then, we claim that if we can improve over the iterations the MPC model, that is, learn over iterations the uncertainties, then we can improve over time the MPC performance, either in terms of stabilization or tracking. Before formulating this idea algorithmically, we recall the principle of model-free dither-based MES control. To use the dither-based MES learning algorithm, we define the learning cost function as

$$Q(\hat{\Delta}) = F(\gamma_e(\hat{\Delta})), \qquad (6.17)$$

where $\gamma_e = \gamma - \gamma_{\text{ref}}$ is a tracking error between the system output γ, and a desired output reference γ_{ref}. $\hat{\Delta}$ is the vector obtained by concatenating all the elements of the estimated uncertainty matrices $\Delta\hat{A}$, $\Delta\hat{B}$, $\Delta\hat{C}$, and $\Delta\hat{D}$, $F: \mathbb{R}^p \to \mathbb{R}$, $F(0) = 0$, $F(\gamma_e) > 0$ for $\gamma_e \neq 0$.

In order to ensure convergence of the dither-based MES algorithm, the learning cost function Q needs to satisfy the following assumptions.

Assumption 6.11. *The cost function Q has a local minimum at $\hat{\Delta}^* = \Delta$.*

Assumption 6.12. *The original value of the parameters' estimates $\hat{\Delta}$ is close enough to the actual parameters' values Δ.*

Assumption 6.13. *The cost function is analytic and its variation with respect to the uncertain variables is bounded in the neighborhood of Δ^*, that is, there exists $\xi_2 > 0$, such that $\|\frac{\partial Q}{\partial \Delta}(\tilde{\Delta})\| \leq \xi_2$ for all $\tilde{\Delta} \in \mathcal{V}(\Delta^*)$, where $\mathcal{V}(\Delta^*)$ denotes a compact neighborhood of Δ^*.*

Remark 6.11. *We wrote the cost function (6.17) as a function of the tracking error γ_e; however, the case of regulation or stabilization can be readily deduced from this formulation by replacing the time-varying reference with a constant reference or an equilibrium point.*

Under Assumptions 6.11–6.13, it has been shown (e.g., Ariyur and Krstic, 2002; Nesic, 2009) that the following dither-based MES converges to the local minima of Q:

$$
\begin{aligned}
\dot{z}_i &= a_i \sin\left(\omega_i t + \frac{\pi}{2}\right) Q(\hat{\Delta}), \\
\hat{\Delta}_i &= z_i + a_i \sin\left(\omega_i t - \frac{\pi}{2}\right), \quad i \in \{1, \ldots, N_p\}
\end{aligned}
\tag{6.18}
$$

where $N_p \leq nn + nm + pn + pm$ is the number of uncertain elements, $\omega_i \neq \omega_j, \omega_i + \omega_j \neq \omega_k, i, j, k \in \{1, \ldots, N_p\}$, and $\omega_i > \omega^*, \forall i \in \{1, \ldots, N_p\}$, with ω^* large enough.

The idea that we want to use here is that under Assumptions 6.11–6.13 we can merge an MPC algorithm and a discrete-time version of the MES algorithm to obtain an ESILC-MPC algorithm. We formalize this idea in the iterative Algorithm 6.2.

6.5.3 Stability Discussion

As mentioned earlier, we do not aim in this second part of the chapter to have rigorous guarantees on the MPC robust feasibility. The goal here is to present a simple way of implementing an iterative ES-based MPC. However, we want to sketch an approach to analyze the stability of Algorithm 6.2, if one wanted to add more constraints on the MPC problem to ensure some sort of input-to-state boundedness. By Assumptions 6.8–6.10, the model structural uncertainties $\Delta A, \Delta B, \Delta C$, and ΔD are bounded, the uncertain model matrices $A + \Delta A$, $B + \Delta B$, $C + \Delta C$, and $D + \Delta D$ are elements of convex sets $\mathcal{K}_a, \mathcal{K}_b, \mathcal{K}_c$, and $\mathcal{K}_d$, and finally, the MPC problem (6.19) is well posed. Based on this, the approach for proving stability is based on establishing a boundedness of the tracking error norm $\|y_e\|$ with the upper bound being a function of the uncertainties' estimation error norm $\|\hat{\Delta} - \Delta\|$. One effective way to characterize such a bound is to use ISS between the input $\|\hat{\Delta} - \Delta\|$ and the augmented state $\|y_e\|$. If the ISS property is obtained, then by reducing the estimation error $\|\hat{\Delta} - \Delta\|$ we also reduce the tracking error $\|y_e\|$, due to the ISS relation between the two signals. Based on Assumptions 6.11–6.13, it is known (e.g., Ariyur and Krstic, 2002; Nesic, 2009) that the dither-based MES algorithm (6.20) converges to a local minimum of the learning cost Q, which implies (based on Assumption 6.13) that the estimation error $\|\hat{\Delta} - \Delta\|$ is reducing over the dither-based MES iterations. Thus, finally we conclude that the MPC tracking (or regulation) performance is improved over the iterations of Algorithm 6.2.

ALGORITHM 6.2   Dither-Based ESILC-MPC

- Initialize $z_i(0) = 0$, and the uncertainties' vector estimate $\hat{\Delta}(0) = 0$.
- Choose a threshold for the cost function minimization $\epsilon_Q > 0$.
- Choose the parameters, MPC sampling time $\delta T_{mpc} > 0$, and MES sampling time $\delta T_{mes} = N_E \delta T_{mpc}$, $N_E > 0$.
- Choose the MES dither signals' amplitudes and frequencies: $a_i, \omega_i, i = 1, 2 \ldots, N_p$.

WHILE(true)

    FOR($\ell = 1, \ell \leq N_E, \ell = \ell + 1$)

- Solve the MPC problem

$$\min_{U(k)} \sum_{i=0}^{N-1} \|x(i|k)\|_{QM}^2 + \|u(i|k)\|_{RM}^2 + \|x(N|k)\|_{PM}^2, \quad (6.19a)$$

$$\text{s.t.} \quad x(i+1|k) = (A + \Delta A)x(i|k) + (B + \Delta B)u(i|k),$$

$$y(i|k) = (C + \Delta C)x(i|k) + (D + \Delta D)u(i|k),$$

$$x_{min} \leq x(i|k) \leq x_{max}, \quad i \in \mathbb{Z}_{[1,N_c]}, \quad (6.19b)$$

$$u_{min} \leq u(i|k) \leq u_{max}, \quad i \in \mathbb{Z}_{[0,N_{cu}-1]}, \quad (6.19c)$$

$$y_{min} \leq y(i|k) \leq y_{max}, \quad i \in \mathbb{Z}_{[1,N_c]}, \quad (6.19d)$$

$$x(0|k) = x(k), \quad (6.19e)$$

- Update $k = k + 1$.

    End

      IF $Q > \epsilon_Q$

- Evaluate the MES cost function $Q(\hat{\Delta})$
- Evaluate the new value of the uncertainties $\hat{\Delta}$:

$$z_i(h+1) = z_i(h) + a_i \delta T_{mes} \sin\left(\omega_i h \delta T_{mes} + \frac{\pi}{2}\right) Q(\hat{\Delta}),$$

$$\hat{\Delta}_i(h+1) = z_i(h+1) + a_i \sin\left(\omega_i h \delta T_{mes} - \frac{\pi}{2}\right), \quad (6.20)$$

$$i \in \{1, \ldots, N_p\}$$

- Update $h = h + 1$.

    End

    Reset $\ell = 0$

End

6.6 NUMERICAL EXAMPLES

6.6.1 Example for the DIRECT-Based ILC MPC

In this section we present some numerical results for the DIRECT ILC-based MPC scheme. We consider the following simple system dynamics:

$$x_1(k+1) = x_1(k) + (-1 + 3/(k_1 + 1))x_2(k) + u(k),$$
$$x_2(k+1) = -k_2 x_2(k) + u(k),$$
$$y(k) = x_1(k)$$

together with the state constraints $|x_i| \leq 50$ for $i \in \{1, 2\}$, and the input constraint $|u| \leq 10$. We assume that the nominal values for the parameters are $k_1 = -0.3$ and $k_2 = 1$. Furthermore, we assume that the uncertainties are bounded by $-0.9 \leq k_1 \leq 0$ and $0 \leq k_2 \leq 4$. The robust invariant sets for tracking are determined using the algorithms in Borrelli et al. (2012).

Next, we define an identification-type cost function (see Chapter 5): for a given initial condition x_0 and a piecewise constant reference trajectory of length T sufficiently large, the learning cost function Q is defined as

$$Q(\hat{\Delta}) := \sum_{k=1}^{T} \|x(k) - \tilde{x}(k)\|^2,$$

where $x(k)$ is the trajectory of the actual system (assuming the availability of the full states' measurements), and $\tilde{x}(k)$ is the trajectory of the MPC model. Note that the trajectory $\tilde{x}(k)$ is generated by applying the first input of the optimal control sequence, which is the same as the MPC law. Hence, we apply the same input (MPC law) to both the system and the MPC model to calculate the cost function. Without loss of generality if we start from x_0, and the control at each stage is given by $\kappa_N(x(k))$, we can write the following states' evolutions:

$$x(k+1) = F(x(k), \kappa_N(x(k)), \hat{\Delta}) + w(k),$$
$$\tilde{x}(k+1) = F(\tilde{x}(k), \kappa_N(x(k)), \hat{\Delta}),$$
$$x(0) = \tilde{x}(0) = x_0.$$

This choice of the learning cost function Q (an identification–type cost function) differs from conventional ILC methods which use an error between a desired reference trajectory and the actual trajectory to compute the learning cost function. We will use such a kind of cost function in this example.

In this case, it can be easily observed that for every $k > 0$,

$$Q_k(\hat{\Delta}) := \|x(k) - \tilde{x}(k)\|^2 = \|(\hat{\Delta}A)^{k-1}w(0) + (\hat{\Delta}A)^{k-2}w(1)$$
$$+ \cdots + w(k-1)\|^2.$$

From the structure of $w(k)$, it follows that $Q(\hat{\Delta}) = 0$ if $\hat{\Delta} = \Delta$. Because Q is a positive function it follows that Q has a global minimum at Δ. It might be required to explicitly know the control law to guarantee that Δ is the unique minimum. Although in this example we can observe from the plot of Q in Fig. 6.1 that the unique minimum occurs at the true value of the parameters. The smoothness of Q depends on how the parameter Δ affects the system matrices and the control law. Fig. 6.3 shows the identification of the uncertainties obtained by the DIRECT algorithm. Finally, we see in Fig. 6.2 that the ESILC-MPC scheme successfully achieves tracking of the piecewise constant reference trajectory.

6.6.2 Example for the Dither-Based ESILC-MPC

We report here some numerical results on the direct current servomotor. We consider the case where the motor is connected via a flexible shaft to a load. The goal is to control the load's angular position. The states of

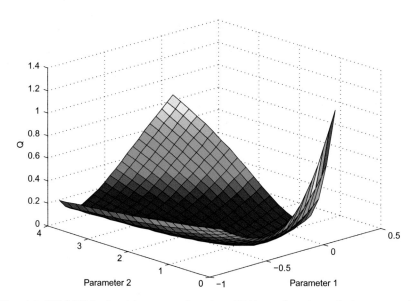

Fig. 6.1 ESILC-MPC algorithm: cost function $Q(\Delta)$ as function of the uncertain parameters.

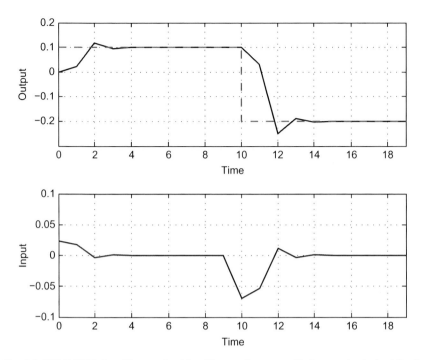

Fig. 6.2 ESILC-MPC algorithm: output tracking performance (Reference in dashed-line).

the system are the load angle, the load angular rate, the motor angle, and the motor angular rate. The control input is the motor voltage, and the controlled outputs are the load angle and the torque acting on the flexible shaft. The model for the system in state space can be written as (e.g., see Benosman et al., 2014 and references therein)

$$
\dot{x}_c(t) = \begin{bmatrix} 0 & 1 & 0 & 0 \\ -\dfrac{k_l}{\mathcal{J}_l} & -\dfrac{\beta_l}{\mathcal{J}_l} & \dfrac{k_l}{g\mathcal{J}_l} & 0 \\ 0 & 0 & 0 & 1 \\ \dfrac{k_l}{g\mathcal{J}_m} & 0 & -\dfrac{k_l}{g^2\mathcal{J}_m} & -\dfrac{\beta_m+R_A^{-1}K_m^2}{\mathcal{J}_m} \end{bmatrix} x_c(t) + \begin{bmatrix} 0 \\ 0 \\ 0 \\ \dfrac{K_m}{R_A\mathcal{J}_m} \end{bmatrix} u_c(t),
$$

$$
y_c(t) = \begin{bmatrix} 1 & 0 & 0 & 0 \\ k_l & 0 & -\dfrac{k_l}{g} & 0 \end{bmatrix} x_c(t),
$$

$$(6.21)$$

where $x_c \in \mathbb{R}^4$ is the state vector, $u_c \in \mathbb{R}$ is the input vector, and $y_c \in \mathbb{R}^2$ is the output vector. In Eq. (6.21) $\mathcal{J}_l$ [kg m^2], β_l [Nms/rad],

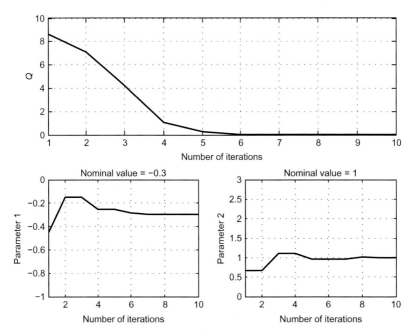

Fig. 6.3 ESILC-MPC algorithm: learning cost function and uncertainties estimation over learning iterations.

and k_l [Nm/rad] are the inertia, friction, and stiffness of the load and the flexible shaft, respectively. R_A [Ω] is the armature resistance, K_m [Nm/A] is the motor constant, $\mathcal{J}_m$ [kg m^2], β_m [Nms/rad] are the inertia and friction of the motor, respectively. g is the gear ratio between the motor and the load. The nominal numerical values used in the simulations are $R_A = 20$ Ω, $K_m = 15$ Nm/A, $\mathcal{J}_l = 45$ kg m^2, $\beta_l = 45$ Nms/rad, $k_l = 1.92 \times 10^3$ Nm/rad, $\mathcal{J}_m = 0.5$ kg m^2, and $\beta_m = 0.1$ Nms/rad.

The system is subject to constraints on the motor voltage and the shaft torque:

$$-78.5 \leq y_{c-2}(t) \leq 78.5, \tag{6.22a}$$

$$-220 \leq u_c(t) \leq 220. \tag{6.22b}$$

The control objective is to track a time-varying load angle position reference signal $r_l(t)$. Following the nominal MPC presented in Section 6.5.1, the prediction model is obtained by sampling Eq. (6.21) with a period

$\delta T_{\mathrm{mpc}} = 0.1$ s. A classical formulation of the control input is used:

$$u_c(k + 1) = u_c(k) + \Delta v(k).$$

Next, the MPC cost function is chosen as

$$\sum_{i=1}^{N} \| y_{c-1}(i|k) - \eta(i|k) \|_{Q_y}^2 + \| \Delta v(i|k) \|_{R_v}^2 + \rho \sigma^2, \qquad (6.23)$$

where $Q_y = 10^3$ and $R_v = 0.05$, prediction, constraints, and control horizons are $N = 20$, $N_c = N_{cu} = N_u = 4$. In this case the output constraints (6.22a) are considered as soft constraints, which could be (briefly) violated due to the mismatch between the prediction model and the actual model; that is, the robust MPC feasibility is not guaranteed as in the DIRECT-based ESILC-MPC. Thus, in Eq. (6.23) we add the term $\rho \sigma^2$ where $\rho > 0$ is a (large) cost weight, and σ is an additional variable used to model the (maximum) constraint violation of the softened constraints. We consider the initial state $x(0) = [0\ 0\ 0\ 0]'$ and the reference $\eta(t) = 4.5 \sin(\frac{2\pi}{T_{\mathrm{ref}}} t)$, with $T_{\mathrm{ref}} = 20\pi$ s.

First, to verify the baseline performance, we solve the nominal MPC problem, that is, without model uncertainties.[2] We report the corresponding results in Fig. 6.4, where it is clear that the desired load angular position

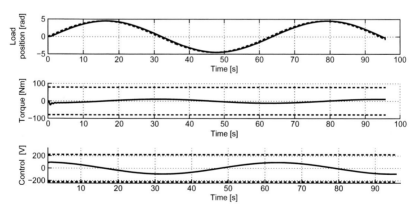

Fig. 6.4 Output and input signals in the nominal case (reference trajectory and constraints in *dashed line*, obtained signals in *solid line*).

---

[2]Such basic MPC problems for linear time–invariant systems can be readily solved using, for example, the free Multi-Parametric Toolbox (MPT 2.6).

is precisely tracked, without violating the problem constraints. Next, we introduce the parametric model uncertainties $\delta\beta_l = -80$ Nms/rad, $\delta k_l = -100$ Nm/rad. Note that we purposely introduce large parametric uncertainties to show clearly the bad effect of these uncertainties on the nominal MPC algorithm, and to subsequently test the dither-based ESILC-MPC algorithm on this challenging case.

We first apply the nominal MPC controller to the uncertain model; we show the obtained performance in Fig. 6.5, where it is clear that the nominal performance is lost.

Now, we apply the ESILC-MPC Algorithm 6.2, where we set $\delta T_{mes} = 1.5 T_{ref}$. We choose the MES learning cost function as

$$Q = \sum_{i=0}^{N_E-1} \|y_e(i\delta T_{mpc})\|^2 + \|\dot{y}_e(i\delta T_{mpc})\|^2, \, y_e = y_{c-1} - r_l,$$

that is, the norm of the error in the load angular position and velocity. To learn the uncertain parameter β_l we apply algorithm (6.20), as follows:

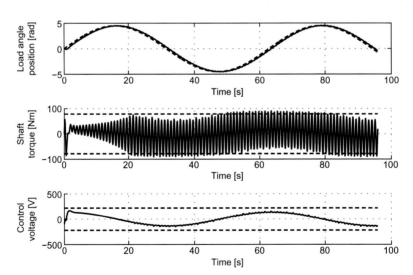

Fig. 6.5 Output and input signals in the uncertain case with nominal MPC (reference trajectory and constraints in *dashed line*, obtained signals in *solid line*).

$$z_{\beta_l}(k'+1) = z_{\beta_l}(k') + a_{\beta_l} \delta T_{\text{mes}} \sin\left(\omega_{\beta_l} k' \delta T_{\text{mes}} + \frac{\pi}{2}\right) Q,$$
$$\delta\hat{\beta}_l(k'+1) = z_{\beta_l}(k'+1) + a_{\beta_l} \sin\left(\omega_{\beta_l} k' \delta T_{\text{mes}} - \frac{\pi}{2}\right),$$
(6.24)

with $a_{\beta_l} = 2 \times 10^{-7}, \omega_{\beta_l} = 0.7$ rad/s.

To learn the uncertain parameter k_l, we apply algorithm (6.20), as

$$z_{k_l}(k'+1) = z_{k_l}(k') + a_{k_l} \delta T_{\text{mes}} \sin\left(\omega_{k_l} k' \delta T_{\text{mes}} + \frac{\pi}{2}\right) Q,$$
$$\delta\hat{k}_l(k'+1) = z_{k_l}(k'+1) + a_{k_l} \sin\left(\omega_{k_l} k' \delta T_{\text{mes}} - \frac{\pi}{2}\right),$$
(6.25)

with $a_{k_l} = 3.8 \times 10^{-7}, \omega_{k_l} = 1$ rad/s.

We select $\omega_{\beta_l}, \omega_{k_l}$ to be higher than the desired frequency of the closed loop (around 0.1 rad/s) to ensure convergence of the MES algorithm because the ES algorithms' convergence proofs are based on averaging theory, which assumes high dither frequencies (e.g., Ariyur and Krstic, 2002; Nesic, 2009). We set the MES cost function threshold ϵ_Q to $1.5 Q_{\text{nominal}}$, where Q_{nominal} is the value of the MES cost function obtained in the nominal-model case with the nominal MPC (i.e., the baseline ideal case). In other words, we decide to stop searching for the best estimation of the value of the uncertainty when the uncertain MES cost function, that is, the value of Q when applying the ESILC-MPC algorithm to the uncertain model is less than or equal to 1.5 of the MES cost function in the case without model uncertainties, which represents the best achievable MES cost function.

The obtained results with the dither-based ESILC-MPC algorithm are reported in Figs. 6.6 to 6.9. First, note that in Fig. 6.6 the learning cost function decreases as expected along the MES learning iterations to reach a small value after only few iterations. This corresponds to the required number of iterations to learn the actual value of the uncertain parameters as shown in Figs. 6.7 and 6.8. Eventually, after the convergence of the learning algorithm, we see in Fig. 6.9 that the nominal baseline performances of the MPC are recovered and that the output tracks the desired reference without violating the desired constraints.

6.7 CONCLUSION AND OPEN PROBLEMS

In this chapter, we have reported some results about ES-based adaptive MPC algorithms. We have argued that it is possible to merge together

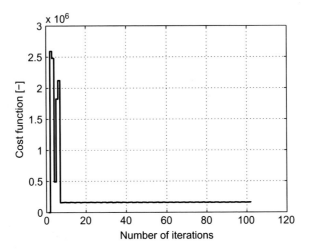

Fig. 6.6 MES cost function evolution over the ESILC-MPC learning iterations.

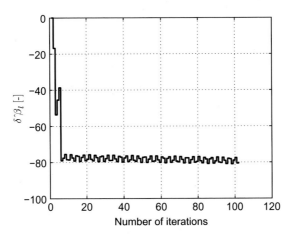

Fig. 6.7 Uncertain parameter $\delta \beta_l$ learning evolution over the ESILC-MPC learning iterations.

a model-based linear MPC algorithm with a model-free ES algorithm to iteratively learn structural model uncertainties and thus improve the overall performance of the MPC controller. We have presented the stability analysis of this modular design technique for learning-based adaptive MPC. The approach taken in this chapter carefully addresses both feasibility and tracking performance for the ESILC-MPC scheme. Future work can include extending this method to a wider class of nonlinear systems, tracking a richer class of signals, and employing different nonsmooth

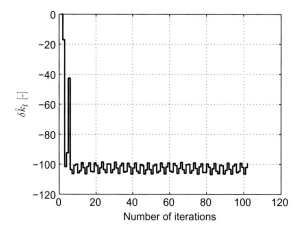

Fig. 6.8 Uncertain parameter δk_l learning evolution over the ESILC-MPC learning iterations.

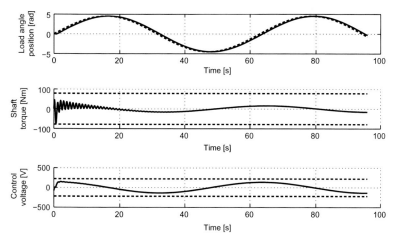

Fig. 6.9 Output and input signals in the uncertain case with the ESILC-MPC (reference trajectory and constraints in *dashed line*, obtained signals in *solid line*).

optimization techniques for the ES algorithm. Other research directions could consider improving the convergence rate of the iterative learning MPC algorithm by using different ES algorithms with semiglobal convergence properties (e.g., Tan et al., 2006; Noase et al., 2011); extending this work to different types of model-free learning algorithms (e.g., reinforcement learning algorithms); and comparing the obtained ILC-MPC algorithms in terms of their convergence rate and optimal performances.

REFERENCES

Adetola, V., DeHaan, D., Guay, M., 2009. Adaptive model predictive control for constrained nonlinear systems. Syst. Control Lett. 58 (5), 320–326.

Ahn, H.S., Chen, Y., Moore, K.L., 2007. Iterative learning control: brief survey and categorization. IEEE Trans. Syst. Man Cybern. 37 (6), 1099.

Alvarado, I., Limon, D., Alamo, T., Camacho, E., 2007a. Output feedback robust tube based MPC for tracking of piece-wise constant references. In: Proceedings of the 46th IEEE Conference on Decision and Control. IEEE, pp. 2175–2180.

Alvarado, I., Limon, D., Alamo, T., Fiacchini, M., Camacho, E., 2007b. Robust tube based MPC for tracking of piece-wise constant references. In: 46th IEEE Conference on Decision and Control. IEEE, pp. 1820–1825.

Arimoto, S., 1990. Robustness of learning control for robot manipulators. In: Proceedings of the IEEE International Conference on Robotics and Automation. IEEE, pp. 1528–1533.

Ariyur, K.B., Krstic, M., 2002. Multivariable extremum seeking feedback: analysis and design. In: Proceedings of the Mathematical Theory of Networks and Systems, South Bend, IN.

Ariyur, K.B., Krstic, M., 2003. Real-Time Optimization by Extremum-Seeking Control. John Wiley & Sons, New York, NY, USA.

Aswani, A., Bouffard, P., Tomlin, C., 2012a. Extensions of learning-based model predictive control for real-time application to a quadrotor helicopter. In: American Control Conference (ACC), 2012. IEEE, pp. 4661–4666.

Aswani, A., Master, N., Taneja, J., Culler, D., Tomlin, C., 2012b. Reducing transient and steady state electricity consumption in HVAC using learning-based model-predictive control. Proc. IEEE 100 (1), 240–253.

Aswani, A., Gonzalez, H., Sastry, S.S., Tomlin, C., 2013. Provably safe and robust learning-based model predictive control. Automatica 49 (5), 1216–1226.

Bemporad, A., 1998. Reference governor for constrained nonlinear systems. IEEE Trans. Autom. Control 43 (3), 415–419.

Bemporad, A., Morari, M., Dua, V., Pistikopoulos, E.N., 2002. The explicit linear quadratic regulator for constrained systems. Automatica 38 (1), 3–20.

Benosman, M., Atinc, G., 2013a. Multi-parametric extremum seeking-based learning control for electromagnetic actuators. In: IEEE, American Control Conference, Washington, DC, pp. 1914–1919.

Benosman, M., Atinc, G., 2013b. Nonlinear learning-based adaptive control for electromagnetic actuators. In: IEEE, European Control Conference, Zurich, pp. 2904–2909.

Benosman, M., Lum, K.Y., 2009. Online References Reshaping and Control Reallocation for Nonlinear Fault Tolerant Control. IEEE Trans. Control Syst. Technol. 17 (2), 366–379.

Benosman, M., Di Cairano, S., Weiss, A., 2014. Extremum seeking-based iterative learning linear MPC. In: IEEE Multi-conference on Systems and Control, pp. 1849–1854.

Borrelli, F., Bemporad, A., Morari, M., 2012. Predictive control. Draft Available at: http://www.mpc.berkeley.edu/mpc-course-material.

Cai, C., Teel, A.R., 2009. Characterizations of input-to-state stability for hybrid systems. Syst. Control Lett. 58 (1), 47–53.

Cueli, J.R., Bordons, C., 2008. Iterative nonlinear model predictive control. Stability, robustness and applications. Control Eng. Pract. 16 (9), 1023–1034.

Ferramosca, A., Limón, D., Alvarado, I., Alamo, T., Camacho, E. F., 2009. MPC for tracking with optimal closed-loop performance. Automatica 45 (8), 1975–1978.

Finkel, D.E., 2003. DIRECT Optimization Algorithm User Guide, vol. 2. Center for Research in Scientific Computation, North Carolina State University.

Genceli, H., Nikolaou, M., 1996. New approach to constrained predictive control with simultaneous model identification. AIChE J. 42 (10), 2857–2868.

González, A., Ferramosca, A., Bustos, G., Marchetti, J., Odloak, D., 2013. Model predictive control suitable for closed-loop re-identification. In: IEEE, American Control Conference. IEEE, pp. 1709–1714.

Grancharova, A., Johansen, T., 2009. Explicit model predictive control of an electropneumatic clutch actuator using on/off valves and pulsewidth modulation. In: IEEE, European Control Conference, Budapest, Hungary, pp. 4278–4283.

Guay, M., Dhaliwal, S., Dochain, D., 2013. A time-varying extremum-seeking control approach. In: IEEE, American Control Conference, pp. 2643–2648.

Hartley, E., Jerez, J., Suardi, A., Maciejowski, J., Kerrigan, E., Constantinides, G., 2012. Predictive control of a Boeing 747 aircraft using an FPGA. In: Proceedings of the 4th IFAC Nonlinear Model Predictive Control Conference, Noordwijkerhout, The Netherlands, pp. 80–85.

Heirung, T.A.N., Ydstie, B.E., Foss, B., 2013. An adaptive model predictive dual controller. In: Adaptation and Learning in Control and Signal Processing, vol. 11, pp. 62–67.

Hudon, N., Guay, M., Perrier, M., Dochain, D., 2008. Adaptive extremum-seeking control of convection-reaction distributed reactor with limited actuation. Comput. Chem. Eng. 32 (12), 2994–3001.

Jansson, H., Hjalmarsson, H., 2005. Input design via LMIs admitting frequency-wise model specifications in confidence regions. IEEE Trans. Autom. Control 50 (10), 1534–1549.

Jones, D.R., Perttunen, C.D., Stuckman, B.E., 1993. Lipschitzian optimization without the Lipschitz constant. J. Optim. Theory Appl. 79 (1), 157–181.

Khong, S.Z., Nešić, D., Manzie, C., Tan, Y., 2013a. Multidimensional global extremum seeking via the DIRECT optimisation algorithm. Automatica 49 (7), 1970–1978.

Khong, S.Z., Nešić, D., Tan, Y., Manzie, C., 2013b. Unified frameworks for sampled-data extremum seeking control: global optimisation and multi-unit systems. Automatica 49 (9), 2720–2733.

Krstić, M., 2000. Performance improvement and limitations in extremum seeking control. Syst. Control Lett. 39 (5), 313–326.

Krstić, M., Wang, H.H., 2000. Stability of extremum seeking feedback for general nonlinear dynamic systems. Automatica 36 (4), 595–601.

Lee, K.S., Chin, I.S., Lee, H.J., Lee, J.H., 1999. Model predictive control technique combined with iterative learning for batch processes. AIChE J. 45 (10), 2175–2187.

Limon, D., Alamo, T., Muñoz de la Peña, D., Zeilinger, M.N., Jones, C., Pereira, M., 2012. MPC for tracking periodic reference signals. In: Proceedings of the IFAC Conference on Nonlinear Model Predictive Control, No. EPFL-CONF-181940.

Limón, D., Alvarado, I., Alamo, T., Camacho, E.F., 2008. MPC for tracking piecewise constant references for constrained linear systems. Automatica 44 (9), 2382–2387.

Limon, D., Alvarado, I., Alamo, T., Camacho, E., 2010. Robust tube-based MPC for tracking of constrained linear systems with additive disturbances. J. Process Control 20 (3), 248–260.

Lobo, M.S., Boyd, S., 1999. Policies for simultaneous estimation and optimization. In: IEEE, American Control Conference, pp. 958–964.

Marafioti, G., Bitmead, R.R., Hovd, M., 2014. Persistently exciting model predictive control. Int. J. Adapt. Control Signal Process. 28, 536–552.

Mayne, D.Q., Rawlings, J.B., Rao, C.V., Scokaert, P.O., 2000. Constrained model predictive control: stability and optimality. Automatica 36 (6), 789–814.

Moore, K.L., 1999. Iterative learning control: an expository overview. In: Applied and Computational Control, Signals, and Circuits. Springer, New York, pp. 151–214.

Nesic, D., 2009. Extremum seeking control: convergence analysis. Eur. J. Control 15 (34), 331–347.

Noase, W., Tan, Y., Nesic, D., Manzie, C., 2011. Non-local stability of a multi-variable extremum-seeking scheme. In: IEEE, Australian Control Conference, pp. 38–43.

Popovic, D., Jankovic, M., Magner, S., Teel, A.R., 2006. Extremum seeking methods for optimization of variable cam timing engine operation. IEEE Trans. Control Syst. Technol. 14 (3), 398–407.

Rathouský, J., Havlena, V., 2013. MPC-based approximate dual controller by information matrix maximization. Int. J. Adapt. Control Signal Process. 27 (11), 974–999. http://dx.doi.org/10.1002/acs.2370.

Rotea, M., 2000. Analysis of multivariable extremum seeking algorithms. In: Proceedings of the American Control Conference, vol. 1. IEEE, pp. 433–437.

Shi, J., Gao, F., Wu, T.J., 2007. Single-cycle and multi-cycle generalized 2D model predictive iterative learning control (2D-GPILC) schemes for batch processes. J. Process Control 17 (9), 715–727.

Sotomayor, O.A., Odloak, D., Moro, L.F., 2009. Closed-loop model re-identification of processes under MPC with zone control. Control Eng. Pract. 17 (5), 551–563.

Tan, Y., Nesic, D., Mareels, I., 2006. On non-local stability properties of extremum seeking control. Automatica 42, 889–903.

Teel, A.R., Popovic, D., 2001. Solving smooth and nonsmooth multivariable extremum seeking problems by the methods of nonlinear programming. In: Proceedings of the 2001 American Control Conference, vol. 3. IEEE, pp. 2394–2399.

Wang, Y., Zhou, D., Gao, F., 2008. Iterative learning model predictive control for multi-phase batch processes. J. Process Control 18 (6), 543–557.

Wang, Y., Gao, F., Doyle, F.J., 2009. Survey on iterative learning control, repetitive control, and run-to-run control. J. Process Control 19 (10), 1589–1600.

Žáčeková, E., Prívara, S., Pčolka, M., 2013. Persistent excitation condition within the dual control framework. J. Process Control 23 (9), 1270–1280.

Zhang, C., Ordóñez, R., 2012. Extremum-Seeking Control and Applications. Springer-Verlag, New York.

Zhang, T., Guay, M., Dochain, D., 2003. Adaptive extremum seeking control of continuous stirred-tank bioreactors. AIChE J. 49, 113–123.

CONCLUSIONS AND FURTHER NOTES

We started this journey with the specific idea of presenting our latest results in the field of learning-based adaptive control. These results were obtained over a period of 5 years, and the book project took about 1 year of writing.

Originally, we wanted to focus specifically on learning-based adaptive controllers; however, as the book project advanced, it became clear that this monograph will not be complete without a more general discussion about adaptive control: a discussion in which the reader can see where the learning-based adaptation techniques stand in the bigger picture of adaptive control theory.

To do so, we decided to write Chapter 2, where the aim was to present a recent survey in adaptive control. To be honest, that chapter was the hardest to "finalize." The reason is that the number of papers, reports, and monographs on adaptive control is staggering. Indeed, if we were to write a full survey of all the results, we would have to dedicate the book in its entirety solely to that purpose. Instead, we considered our goal in writing that chapter in the first place. Because our goal was to present the big picture of adaptive control and situate it in the results of this book, it became clearer to us that one way to achieve this goal was to decompose the entire field of adaptive control into smaller subfields. The way we chose to decompose the field of adaptive control was not the usual way, where the field is decomposed based on the model used to design the controller, that is, linear versus nonlinear, or continuous versus hybrid. Instead, we wanted to talk about what we believe are the three main subfields of adaptive control theory—namely, the subfield of model-based adaptive control (to which we also refer as classical control), the subfield of model-free adaptive control, and the one of interest to us in this monograph: the field of learning-based adaptive control.

Indeed, this decomposition allows us to place the large body of work and results in adaptive control into one of these three subfields. The model-based (or classical) subfield is the one where the system to be controlled is modeled first, and the controller is entirely based on the system's (uncertain) model. We further decomposed this subfield into several subclasses containing the direct adaptive control subclass (based on linear versus nonlinear models) and the indirect adaptive control subclass (based on linear models versus nonlinear models).

To illustrate this subfield we gave a few simple examples in each subclass, and then directed the reader towards more specialized books in each specific subject, for example, for linear model-based adaptive control: Egardt (1979), Landau (1979), Goodwin and Sin (1984), Narendra and Annaswamy (1989), Astrom and Wittenmark (1995), Ioannou and Sun (2012), Landau et al. (2011), and Sastry and Bodson (2011), Tao (2003), and for nonlinear model-based adaptive control: Krstic et al. (1995), Spooner et al. (2002), and Astolfi et al. (2008).

The second subfield was about model-free adaptive controllers. We defined the model-free adaptive control subfield as adaptive methods which do not require any knowledge of the model of the system. Indeed, model-free adaptive controllers learn the optimal or best controller from interaction with the system by trial and error. Examples of such controllers are the neural-network and deep-learning controllers, the pure reinforcement learning (RL) controllers, pure extremum-seeking controllers, and so on. We gave few examples in this subfield, and referred the reader to a more detailed presentation of the field in, for example in, Martinetz and Schulten (1993), Bertsekas and Tsitsiklis (1996), Prabhu and Garg (1996), Sutton and Barto (1998), Ariyur and Krstic (2003), Busonio et al. (2008), Busoniu et al. (2010), Kormushev et al. (2010), Szepesvári (2010), Farahmand (2011), Zhang and Ordóñez (2012), Levine (2013), and Wang et al. (2016).

The last subfield, and the main topic of this book, is learning-based adaptive control. In this class of adaptation, the controller is designed based partially on a model of the system, and partially on a model-free learning algorithm. The idea here is to take advantage of the physics of the system by using a model of the system, albeit incomplete, to design a first "suboptimal" controller, and then complement this controller with a model-free learning algorithm, which is used to compensate for the unknown or uncertain parts of the model. We believe that this type of approach combines the advantages of the two previous subfields. Indeed, this class of adaptive controllers uses the knowledge provided by the physics of the system, and so, unlike the model-free learning algorithms, it does not start learning from scratch. This makes the convergence of the learning-based adaptive algorithms to an optimal controller faster than the convergence of the pure learning algorithms, which have to search over a larger space of controllers. At the same time, learning-based adaptive controllers are more flexible than model-based adaptive controllers in terms of the types of models and uncertainties that they can handle because they do not rely entirely on the model of the system, unlike the model-based controllers. In that sense, we think that this class of adaptive controllers combines the best of both worlds.

In Chapter 2, we also presented some simple examples to illustrate this type of adaptive controllers, and cited some recent literature (including ours) in this line of research, for example, Spooner et al. (2002), Guay and Zhang (2003), Koszaka et al. (2006), Wang et al. (2006), Wang and Hill (2010), Haghi and Ariyur (2011, 2013), Lewis et al. (2012), Zhang and Ordóñez (2012), Benosman and Atinc (2013a,b), Atinc and Benosman (2013), Frihauf et al. (2013), Vrabie et al. (2013), Modares et al. (2013), Benosman et al. (2014), Benosman (2014a,b,c), Benosman and Xia (2015), Gruenwald and Yucelen (2015), and Subbaraman and Benosman (2015). Of course, our own results in this subfield have been presented in more detail in the subsequent chapters of this book, as explained next.

To make the monograph more self-contained, we included in Chapter 1 some definitions and results which are used throughout the book. Indeed, we recalled some basic definitions of vector spaces and set properties as found, for example, in Alfsen (1971), Golub and Van Loan (1996), and Blanchini (1999). We also presented some of the main definitions and theorems in dynamical systems stability, based on the references (Liapounoff, 1949; Zubov, 1964; Rouche et al., 1977; Lyapunov, 1992; Perko, 1996). We finally presented some systems properties such as nonminimum phase property and passivity properties based on Isidori (1989), Ortega et al. (1998), van der Schaft (2000), and Brogliato et al. (2007).

In Chapter 3 we presented our recent results on a specific problem related to learning-based adaptive control, namely, the problem of iterative feedback tuning (IFT). This problem is concerned with online auto-tuning of feedback controllers, based on some measurements from the system. The IFT problem has been largely studied for linear systems, and recently some results have been reported in the more challenging case of nonlinear systems. In this chapter, we decided to focus solely on the nonlinear case. This is, indeed, a characteristic of the results presented in this book, where we chose to focus on nonlinear systems knowing that, with some simplifications, the linear case can usually be recovered from the nonlinear case.

We then decided to study the case of nonlinear models affine in the control vector. This is a general class that has been used to model many practical systems, for example, robotic arms, some unmanned air vehicles, and so on. For this class of nonlinear models, we proposed an IFT approach that, in the spirit of learning-based adaptation, merges together a model-based nonlinear controller and a model-free learning algorithm. The model-based part uses a static-state feedback, complemented with a nonlinear "robustifying" term due to Lyapunov reconstruction techniques. As for the model-free learning algorithm, as the title of the book indicates, it is based on the so-called extremum-seeking algorithms.

We have also reported in Chapter 3 two mechatronics applications of the proposed IFT method. The first application is the control of electromagnetic actuators. These actuators are very useful in industrial applications, for example, electromagnetic brakes, electromagnetic valves, and so on. Furthermore, electromagnetic actuators are often used in cyclic or repetitive tasks, which makes them a very good target for the type of IFT algorithms introduced in this book. The second application is related to the robotics field. We considered the case of rigid manipulator arms. This example is often used as a real-life example in adaptive control due to its popularity as well as its utility. Indeed, a lot of manufacturing relies on robotic arms, and in many cases, the industrial application can change, which implies that the controller designed for a specific application has to be retuned. In this scenario, an auto-tuning, which can be done online without intervention of humans, can indeed be a game changer in terms of efficiency and profitability of the manufacturer. The results reported in Chapter 3 are partly based on the preliminary results published by the author in Benosman and Atinc (2013a) and Benosman (2014c).

Chapter 4 was dedicated to the problem of indirect learning-based adaptive control. Indeed, we considered the problem of controlling nonlinear models with parametric structured uncertainties. As seen in Chapter 2, this problem has been thoroughly studied since the 1950s, but a good universal solution which could work for any type of model nonlinearities and any type of model uncertainties has yet to be found. Many of the available results are restricted to some particular type of nonlinearities and uncertainties. We wanted to present our results in a more general setting of general nonlinear systems with parametric uncertainties. Thus, we first considered the general case of nonlinear models of the form $\dot{x} = f(t, x, u, \Delta)$, where f represents a smooth vector field, and x, u represent the system's states and control, respectively; Δ represents some model parametric uncertainties. We did not impose any explicit limitations of the model structure or the uncertainties structures. For this class of systems, we argued that *if we could* design a feedback controller which renders the closed-loop system input-to-state stable (from a specific input vector to a specific state vector), we could then learn the uncertainties using a model-free learning algorithm, which in our case was an extremum-seeking algorithm.

We then focused on the specific, yet important case of nonlinear models affine in the control, for which we proposed a constructive approach to design a feedback controller that ensures the desired input-to-state stability. This controller was then complemented with a model-free learning algorithm to estimate the uncertainties of the model. We concluded Chapter 4 by studying two mechatronic systems: electromagnetic actuators and rigid manipulator arms. The results reported in Chapter 4 are partly based on the preliminary results published by the author in Atinc and Benosman (2013), Benosman and Atinc (2013b), Benosman (2014a,b), and Benosman and Xia (2015).

Next, in Chapter 5, we studied the problems of real-time system identification and infinite dimension system model reduction. The first problem of system identification is very important is practical applications. Indeed, for many systems, the control engineer knows the general form of the model, usually derived from the laws of physics, but has to calibrate the model by estimating some physical parameters of the system, for example, mass, friction coefficient, and so on. Furthermore, in some cases these parameters are changing over time, either due to the system aging or due to the task undertaken by the system, for example, a robot carrying different payloads over time. For these reasons, identification algorithms are important, especially if they can estimate the model's parameters online while the system is executing its nominal tasks.

For this problem of real-time parametric identification we proposed, for the class of open-loop Lagrange stable systems, an extremum seeking-based identification algorithm that can estimate the system's parameters online. We also considered in Chapter 5 the case of open-loop unstable systems, for which we proposed a closed-loop identification algorithm, the closed-loop being used to stabilize the dynamics, before performing the parametric identification using an extremum seeker. We tested the proposed algorithms on the electromagnetic actuator system, as well as the two-link robot arm example.

The second problem studied in Chapter 5 is the so-called stable model reduction for infinite dimension models. Indeed, infinite dimension models, in the form of partial differential equations (PDEs), are usually obtained from the physics of the system, for example, the wave propagation equation. However, due to their complexity and high dimensionality, they are (besides a few exceptions) difficult to solve in closed form, and their numerical solutions are computationally demanding. To make use of these models in real-time applications, for example, for estimation and control of fluid dynamics, one has first to simplify them to some extent.

The simplification step is sometimes referred to as model reduction because it boils down to reducing the PDE infinite dimensional model to an ordinary differential equation model of finite dimension. However, this step of model reduction may lead, in some cases, to instability (in the Lagrange sense) of the reduced order model (ROM) because the model reduction step involves the removal of some parts of the high order "dynamics" of the PDE model that are sometimes the stabilizing terms of the PDE. To remedy this problem, the model reduction community has investigated the problem of stable model reduction. In this problem, the goal is to design ROMs with solutions replicating (to some extent) the solutions of the full PDE model, while remaining bounded over time. In Chapter 5 we investigated this problem of stable model reduction using the so–called closure models, which are added to the ROMs to stabilize their solutions. We proposed to use model-free extremum seekers to optimally tune the coefficients of some closure models. We tested the proposed stable ROMs' tuning algorithm on the well-known Burgers' equation test-bed. The results reported in Chapter 5 are partly based on the preliminary results published by the author in Benosman et al. (2015b, 2016b).

Finally, in Chapter 6 we studied the problem of modular learning-based adaptive control for linear-invariant systems with structured uncertainties, and under state/input constraints. We studied this problem in the context of model predictive control (MPC). Indeed, MPC is a suitable formulation for control problems, under both state and input constraints. However, one of the main problems that MPC faces in real-life applications is its nonrobustness to model uncertainties. In this context, we used the main idea of this book, combining model-based controllers and a model-free extremum-seeking learning algorithm, to propose an extremum seeking-based iterative learning MPC. Although the theoretical results presented in Chapter 6 are interesting, we believe that their direct implementation in industrial applications is still challenging due to their large computational complexity; for example, the computation in real time of the robust control invariant sets associated with the MPC problem can be challenging. We also presented a simplified version of our extremum seeking-based iterative MPC algorithm that can be easier to implement in real time; however, it does not guarantee that the MPC constraints are satisfied all the time.

We applied this simplified algorithm to a well-known MPC benchmark, namely, the DC servomotor example. The results reported in Chapter 6 are partly based on the preliminary results published by the author in Benosman et al. (2014) and Subbaraman and Benosman (2015).

Besides the results presented in Chapters 3–6 of this monograph, which focused on extremum seeking-based learning adaptive controllers, there are certainly other directions which should be investigated. For instance, because the learning-based adaptive controllers presented here are modular by design, that is, due to their ISS-like properties, one could use other types of model-free learning algorithms to complement the model-based part of the controller. For example, machine learning algorithms, like the RL techniques, could be investigated in this context. By the time we were finalizing this book, we already started investigating the use of some RL algorithms. For instance, in Benosman et al. (2015a, 2016a), we investigated the performances of a Gaussian process upper confidence bound learning algorithm in the context of a modular indirect adaptive control approach.

Finally, to conclude this chapter, and this book, we want to give the reader our "philosophical" opinion about adaptive control in general, and learning-based adaptive control in particular. Indeed, we have been working on adaptive control for about 10 years now, and on learning-based adaptation for about 5 years. Besides our own work in the field, we have read and reviewed a number of papers on adaptive control and learning. Although all of these works solve a specific problem in adaptive control, and definitely advance the field to some extent, none of them is general enough to deal with all real-life situations. Indeed, we saw throughout this book that the modular learning-based design for adaptation can be more general than the "classical" model-based adaptive theory; however, its generality is limited by the learning algorithm itself. What we mean by this is that almost all the available learning algorithms have probes which need to be tuned in such a way that the learning algorithm converges to some reasonable values, leading to some reasonable overall performances of the controlled system. Unfortunately, these tuning parameters of the learning algorithm have to be fine-tuned, which somehow removes all the flexibility introduced by the learning in the first place. With regard to this, we believe that the Holy Grail in adaptation and learning would be a controller which tunes itself, that is, in which its own parameters are auto-tuned. However, it seems utopian to seek such a goal without having to rely on yet another learning algorithm to tune the parameters of the first learning algorithm, and so on. Maybe a more tangible goal would be to seek a learning-based adaptive control where the number of tuning parameters or probes is minimal. Another goal worth targeting would be to seek learning-based adaptive methods which are robust with respect to their tuning parameters choice, that is, which ensure some type of global or semiglobal convergence with respect to their tuning parameters. Maybe such a goal could be attainable by combining tools from model-free learning theory and robust control theory to design robust (with respect to their tuning parameters) learning algorithms. Let us keep looking then!

REFERENCES

Alfsen, E., 1971. Convex Compact Sets and Boundary Integrals. Springer-Verlag, Berlin.

Ariyur, K.B., Krstic, M., 2003. Real Time Optimization by Extremum Seeking Control. John Wiley & Sons, Inc., New York, NY, USA.

Astolfi, A., Karagiannis, D., Ortega, R., 2008. Nonlinear and Adaptive Control with Applications. Springer, London.

Astrom, K.J., Wittenmark, B., 1995. A survey of adaptive control applications. In: IEEE, Conference on Decision and Control, pp. 649–654.

Atinc, G., Benosman, M., 2013. Nonlinear learning-based adaptive control for electromagnetic actuators with proof of stability. In: IEEE, Conference on Decision and Control, Florence, pp. 1277–1282.

Benosman, M., 2014a. Extremum-seeking based adaptive control for nonlinear systems. In: IFAC World Congress, Cape Town, South Africa, pp. 401–406.

Benosman, M., 2014b. Learning-based adaptive control for nonlinear systems. In: IEEE European Control Conference, Strasbourg, FR, pp. 920–925.

Benosman, M., 2014c. Multi-parametric extremum seeking-based auto-tuning for robust input-output linearization control. In: IEEE, Conference on Decision and Control, Los Angeles, CA, pp. 2685–2690.

Benosman, M., Atinc, G., 2013a. Multi-parametric extremum seeking-based learning control for electromagnetic actuators. In: IEEE, American Control Conference, Washington, DC, pp. 1914–1919.

Benosman, M., Atinc, G., 2013b. Nonlinear learning-based adaptive control for electromagnetic actuators. In: IEEE, European Control Conference, Zurich, pp. 2904–2909.

Benosman, M., Xia, M., 2015. Extremum seeking-based indirect adaptive control for nonlinear systems with time-varying uncertainties. In: IEEE, European Control Conference, Linz, Austria, pp. 2780–2785.

Benosman, M., Cairano, S.D., Weiss, A., 2014. Extremum seeking-based iterative learning linear MPC. In: IEEE Multi-Conference on Systems and Control, pp. 1849–1854.

Benosman, M., Farahmand, A.M., Xia, M., 2015a. Learning-based modular indirect adaptive control for a class of nonlinear systems, Tech. rep., arXiv:1509.07860 [cs.SY].

Benosman, M., Kramer, B., Boufounos, P., Grover, P., 2015b. Learning-based reduced order model stabilization for partial differential equations: application to the coupled burgers equation, Tech. rep., arXiv:1510.01728 [cs.SY].

Benosman, M., Farahmand, A.M., Xia, M., 2016a, Learning-based modular indirect adaptive control for a class of nonlinear systems. In: IEEE, American Control Conference (in press).

Benosman, M., Kramer, B., Boufounos, P., Grover, P., 2016b, Learning-based modular indirect adaptive control for a class of nonlinear systems. In: IEEE, American Control Conference (in press).

Bertsekas, D., Tsitsiklis, J., 1996. Neurodynamic Programming. Athena Scientific, Cambridge, MA.

Blanchini, F., 1999. Set invariance in control—a survey. Automatica 35 (11), 1747–1768.

Brogliato, B., Lozano, R., Mashke, B., Egeland, O., 2007. Dissipative Systems Analysis and Control. Springer-Verlag, Great Britain.

Busonio, L., Babuska, R., Schutter, B.D., 2008. A comprehensive survey of multiagent reinforcement learning. IEEE Trans. Syst. Man Cybern. C: Appl. Rev. 38 (2), 156–172.

Busoniu, L., Babuska, R., De Schutter, B., Ernst, D., 2010. Reinforcement learning and dynamic programming using function approximators, Automation and Control Engineering. CRC Press, Boca Raton, FL.

Egardt, B., 1979. Stability of Adaptive Controllers. Springer-Verlag, Berlin.

Farahmand, A.M., 2011. Regularization in reinforcement learning. Ph.D. Thesis, University of Alberta.

Frihauf, P., Krstic, M., Basar, T., 2013. Finite-horizon LQ control for unknown discrete-time linear systems via extremum seeking. Eur. J. Control 19 (5), 399–407.

Golub, G., Van Loan, C., 1996. Matrix Computations, third ed. The Johns Hopkins University Press, Baltimore, MD.

Goodwin, G.C., Sin, K.S., 1984. Adaptive Filtering Prediction and Control. Prentice-Hall, Englewood Cliffs, NJ.

Gruenwald, B., Yucelen, T., 2015. On transient performance improvement of adaptive control architectures. Int. J. Control 88 (11), 2305–2315.

Guay, M., Zhang, T., 2003. Adaptive extremum seeking control of nonlinear dynamic systems with parametric uncertainties. Automatica 39, 1283–1293.

Haghi, P., Ariyur, K., 2011. On the extremum seeking of model reference adaptive control in higher-dimensional systems. In: IEEE, American Control Conference, pp. 1176–1181.

Haghi, P., Ariyur, K., 2013. Adaptive feedback linearization of nonlinear MIMO systems using ES-MRAC. In: IEEE, American Control Conference, pp. 1828–1833.

Ioannou, P., Sun, J., 2012. Robust Adaptive Control. Dover Publications, Mineola, NY.

Isidori, A., 1989. Nonlinear Control Systems, second ed. Communications and Control Engineering Series. Springer-Verlag, London.

Kormushev, P., Calinon, S., Caldwell, D.G., 2010. Robot motor skill coordination with EM-based reinforcement learning. In: IEEE/RSJ International Conference on Intelligent Robots and Systems, Taipei, Taiwan, pp. 3232–3237.

Koszaka, L., Rudek, R., Pozniak-Koszalka, I., 2006. An idea of using reinforcement learning in adaptive control systems. In: International Conference on Networking, International Conference on Systems and International Conference on Mobile Communications and Learning Technologies, 2006. ICN/ICONS/MCL 2006, p. 190.

Krstic, M., Kanellakopoulos, I., Kokotovic, P., 1995. Nonlinear and Adaptive Control Design. John Wiley & Sons, New York.

Landau, I.D., 1979. Adaptive Control. Marcel Dekker, New York.

Landau, I.D., Lozano, R., M'Saad, M., Karimi, A., 2011. Adaptive Control: Algorithms, Analysis and Applications, Communications and Control Engineering. Springer-Verlag, London.

Levine, S., 2013. Exploring deep and recurrent architectures for optimal control. In: Neural Information Processing Systems (NIPS) Workshop on Deep Learning.

Lewis, F.L., Vrabie, D., Vamvoudakis, K.G., 2012. Reinforcement learning and feedback control: using natural decision methods to design optimal adaptive controllers. IEEE Control. Syst. Mag. 76–105, http://dx.doi.org/10.1109/MCS.2012.2214134.

Liapounoff, M., 1949. Problème Général de la Stabilité du Mouvement. Princeton University Press, Princeton, NJ, tradui du Russe (M. Liapounoff, 1892, Société mathématique de Kharkow) par M. Édouard Davaux, Ingénieur de la Marine à Toulon.

Lyapunov, A., 1992. The General Problem of the Stability of Motion. Taylor & Francis, Great Britain, with a biography of Lyapunov by V.I. Smirnov and a bibliography of Lyapunov's works by J.F. Barrett.

Martinetz, T., Schulten, K., 1993. A neural network for robot control: cooperation between neural units as a requirement for learning. Comput. Electr. Eng. 19 (4), 315–332.

Modares, R., Lewis, F., Yucelen, T., Chowdhary, G., 2013. Adaptive optimal control of partially-unknown constrained-input systems using policy iteration with experience replay. In: AIAA Guidance, Navigation, and Control Conference, Boston, MA, http://dx.doi.org/10.2514/6.2013-4519.

Narendra, K.S., Annaswamy, A.M., 1989. Stable Adaptive Systems. Prentice-Hall, Englewood Cliffs, NJ.

Ortega, R., Loria, A., Nicklasson, P., Sira-Ramirez, H., 1998. Passivity-Based Control of Euler-Lagrange Systems. Springer-Verlag, Great Britain.

Perko, L., 1996. Differential Equations and Dynamical Systems, Texts in Applied Mathematics. Springer, New York.

Prabhu, S.M., Garg, D.P., 1996. Artificial neural network based robot control: an overview. J. Intell. Robot. Syst. 15 (4), 333–365.

Rouche, N., Habets, P., Laloy, M., 1977. Stability theory by Liapunov's direct method, Applied Mathematical Sciences, vol. 22. Springer-Verlag, New York.

Sastry, S., Bodson, M., 2011. Adaptive Control: Stability, Convergence and Robustness. Dover Publications, Mineola.

Spooner, J.T., Maggiore, M., Ordonez, R., Passino, K.M., 2002. Stable adaptive control and estimation for nonlinear systems. Wiley-Interscience, New York.

Subbaraman, A., Benosman, M., 2015. Extremum seeking-based iterative learning model predictive control (ESILC-MPC), Tech. rep., arXiv:1512.02627v1 [cs.SY].

Sutton, R.S., Barto, A.G., 1998. Reinforcement Learning: An Introduction. MIT Press, Cambridge, MA.

Szepesvári, C., 2010. Algorithms for Reinforcement Learning. Morgan & Claypool Publishers, California, USA.

Tao, G., 2003. Adaptive Control Design and Analysis. John Wiley and Sons, Hoboken, NJ.

van der Schaft, A., 2000. L2-Gain and Passivity Techniques in Nonlinear Control. Springer-Verlag, Great Britain.

Vrabie, D., Vamvoudakis, K., Lewis, F.L., 2013. Optimal Adaptive Control and Differential Games by Reinforcement Learning Principles, IET Digital Library.

Wang, C., Hill, D.J., 2010. Deterministic Learning Theory for Identification, Recognition, and Control. CRC Press, Boca Raton, FL.

Wang, C., Hill, D.J., Ge, S.S., Chen, G., 2006. An ISS-modular approach for adaptive neural control of pure-feedback systems. Automatica 42 (5), 723–731.

Wang, Z., Liu, Z., Zheng, C., 2016. Qualitative analysis and control of complex neural networks with delays, Studies in Systems, Decision and Control, vol. 34. Springer-Verlag, Berlin/Heidelberg.

Zhang, C., Ordóñez, R., 2012. Extremum-Seeking Control and Applications: A Numerical Optimization-Based Approach. Springer-Verlag, London, New York.

Zubov, V., 1964. Methods of A.M. Lyapunov and Their Application. The Pennsylvania State University, State College, PA, translation prepared under the auspices of the United States Atomic Energy Commission.

INDEX

Note: Page numbers followed by *f* indicate figures and *t* indicate tables.

Printed in the United States
By Bookmasters